The MIDI Manual, Second Edition

The MIDI Manual

A Practical Guide to MIDI in the Project Studio

Second Edition

David Miles Huber

Focal Press

Boston   Oxford   Auckland   Johannesburg   Melbourne   New Delhi

Focal Press is an imprint of Butterworth–Heinemann.

Copyright © 1999 by Butterworth–Heinemann

 A member of the Reed Elsevier group

 Recognizing the importance of preserving what has been written, Butterworth–Heinemann prints its books on acid-free paper whenever possible.

 Butterworth–Heinemann supports the efforts of American Forests and the Global ReLeaf program in its campaign for the betterment of trees, forests, and our environment.

Library of Congress Cataloging-in-Publication Data

Huber, David Miles.
    The MIDI manual : a practical guide to MIDI in the project studio
/ David Miles Huber. — 2nd ed.
        p.   cm.
    Includes index.
    ISBN 0-240-80330-2 (pbk. : alk. paper)
    1. MIDI (Standard)   I. Title.
MT723.H82   1999
784.19'0285'46—dc21   98-45592
                                                              CIP

British Library Cataloguing-in-Publication Data
A catalogue record for this book is available from the British Library.

The publisher offers special discounts on bulk orders of this book.
For information, please contact:
Manager of Special Sales
Butterworth–Heinemann
225 Wildwood Avenue
Woburn, MA 01801-2041
Tel: 781-904-2500
Fax: 781-904-2620

For information on all Butterworth–Heinemann publications available, contact our World Wide Web home page at: http://www.bh.com

10   9   8   7   6   5   4   3   2   1

Printed in the United States of America

TABLE OF CONTENTS

6 Editor/Librarians *107*

7 Music Printing Programs *117*

8 Digital Audio in MIDI Production *125*

FOREWORD

The most amazing aspect of the book you're holding in your hands is that it talks about a technological subject that, while more than a decade and a half old, is still vital, growing, and relevant. Think back to other artifacts from the same era as MIDI's early days: Is anyone still writing books about the Commodore 64, the joys of telecommunicating with 300-baud modems, getting the most out of a Mac Plus, or how to organize data on your 20-megabyte hard drive? Of course not. Technology has a certain ruthlessness; in the rush toward the new, the old is forgotten—cast away as an embarrassing reminder of what we once were.

But this hasn't happened with MIDI. Far from disappearing, MIDI has lived long and prospered. Once used by only a handful of computer-literate musicians, MIDI is now commonplace in computers around the world. It has become "the new sheet music," but for computer media rather than the printed page.

What accounts for MIDI's stubborn refusal not to yield to something new and different, and even evolve to newer heights? Part of the reason is that MIDI was born to good parents. Virtually every music-related hardware and software manufacturer in the early 1980s worked together to hammer out a standard: Americans, Japanese, and Europeans debated, refined, and tested until they had forged a truly worldwide phenomenon. They had the good sense to make MIDI inexpensive enough that there was no reason not to include it as part of musical hardware, which created the volume that made it worthwhile for other companies to develop MIDI-compatible widgets.

MIDI was also powerful and useful enough that it served a wide range of musical needs. But the story would have ended there if not for the MIDI Manufacturers Association, the "guardians" of the MIDI spec. Periodically, this organization drafts new proposals that take MIDI to another level. In the years since MIDI's inception, it has become part of studio recording, post-production, video games, lighting control, Broadway shows, the Internet, and much more. All of these improvements have been done in an orderly, consensus-oriented fashion that has guaranteed the spec's universality and effectiveness.

Which brings us to this book. MIDI's evolution has meant that there has been a continuing need for books that explore this evolution. David Huber is an excellent guide to this world. He loves music, technology, and people, which puts him in a good position to communicate about technology to those who want to learn more about this "MIDI thing." Of course, *The MIDI Manual* includes info on MIDI basics, but more importantly, it discusses the many updates and enhancements that have occurred during the past 15 years. While it's a fine place for beginners to get an overview, what's special about *The MIDI Manual* is that it's also "continuing education" for those who want to remain current about MIDI, and get a glimpse of where it's going to be heading in the future.

So sit back, relax, put on a background CD (which, odds are, had MIDI involved somewhere in its production), and find out more about the marvelous world of MIDI. Like a truly fine wine, it doesn't get older—it just gets better.

Craig Anderton

WHAT IS MIDI?

Simply stated, *Musical Instrument Digital Interface (MIDI)* is a digital communications language and compatible hardware specification that allows multiple electronic instruments, performance controllers, computers, and other related devices to communicate with each other over a connected network (Figure 1.1). MIDI is used to translate performance- or control-related events (such as playing a keyboard, selecting a patch number, varying a modulation wheel, etc.) into equivalent digital messages and then transmit these messages to other MIDI devices where they can be used to

Figure 1.1
Example of a typical MIDI system. (Courtesy of Quik Lok, www.musicindustries.com)

control sound generators and performance parameters. The beauty of MIDI is that its data can be recorded into a hardware device or software program (known as a *sequencer*), where it can be edited and transmitted to electronic instruments or other devices to create music.

In artistic terms, this digital language is an important medium that lets artists express themselves with a degree of flexibility and control that was, before its inception, not possible on an individual level. Through the transmission of this performance language, an electronic musician can create and develop a song or composition in a practical, flexible, affordable, and fun production environment.

In addition to composing and performing a song, musicians can also act as techno-conductors, having complete control over a wide palette of sounds, their *timbre* (sound and tonal quality), and overall *blend* (level, panning, and other real-time controls). MIDI can also be used to vary the performance and control parameters of electronic instruments, recording devices, control devices, and signal processors during a performance.

The term *interface* refers to the actual data communications link and hardware in a connected MIDI network. Through the use of MIDI, it's possible for all electronic instruments and devices to be addressed within a network through the transmission of real-time performance and control-related messages. Furthermore, communication with instruments (or individual sound generators in an instrument) can occur using only a single MIDI data line. This is possible because each data line can transmit performance and control messages over 16 discrete channels. This simple fact makes it possible for electronic musicians to record, overdub, mix, and play back their performances in a working environment that loosely resembles the multitrack recording process. Once mastered, however, the repeatability and edit control of MIDI offers production challenges and possibilities that are beyond the capabilities and cost effectiveness of the traditional multitrack recording studio.

A Brief History

In the early days of electronic music, keyboard synthesizers were commonly *monophonic devices* (capable of sounding only one note at a time) and often generated a thin sound quality. These limiting factors caused early manufacturers to look for ways to link more than one instrument together to create a thicker, richer sound texture. This was originally accomplished by establishing an instrument link that would let a synthesizer (acting as a master controller) directly control the performance parameters of one or

control voltage

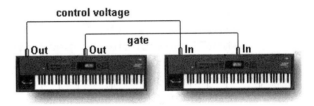

Figure 1.2
Example of two instruments that have been linked using a control voltage and gate signal.

more synthesizers (known as *slave sound modules*). As a result, a basic control signal (known as *control voltage* or Cv) was developed (Figure 1.2).

This simple, yet problematic system was based on the fact that when most early keyboards were played, they generated a DC voltage that could directly control another instrument's voltage-controlled oscillators (which affected the pitch of a sounding note) and voltage-controlled amplifiers (which affected the note's volume and on/off nature). Because many keyboards of the day generated a DC signal that ascended at a rate of 1 volt per octave (breaking each musical octave into 1/12-volt intervals), it was possible to use this standard control voltage as a master-reference signal for transmitting pitch information to other synthesizers. In addition to a control voltage, this standard required that a keyboard transmit a *gate signal*. This second signal was used to synchronize the beginning and duration times of each note. With the appearance of more advanced *polyphonic synthesizers* (which can generate more than one note at a time) and early digital devices, it was clear that this standard would no longer be the answer to system-wide control, and new standards began to appear on the scene (thereby creating incompatible control standards). With the arrival of early drum machines and sequencing devices, standardization became even more of a dilemma.

Synchronization between these early devices was often problematic, because manufacturers would often standardize on different sync-pulse clock rates. Synchronizing incompatible systems was very difficult, because they would lose sync over a very short period of time, rendering synchronization nearly impossible without additional sync-rate converters or other types of modifications.

Because of these incompatibilities, Dave Smith and Chet Wood (then of Sequential Circuits, a now defunct manufacturer of electronic instruments) began creating a digital electronic instrument protocol, which was named the *Universal Synthesizer Interface (USI)*. As a result of this protocol, equipment from different manufacturers could be made to communicate directly with each another. For example, a synthesizer from one company finally worked with another company's sequencer. In the fall of 1981, USI was proposed to the Audio Engineering Society. During the following 2 years, a panel (which included representatives from the major electronic instrument manufacturers) modified this standard and adopted it under the name of Musical Instrument Digital Interface (MIDI Specification 1.0).

The strong acceptance of MIDI was largely due to the need for a standardized protocol and fast-paced advances in technology that allowed complex circuit chips and hardware designs to be manufactured cost effectively. It was also due, in part, to the introduction of Yamaha's popular DX-7 synthesizer in the winter of 1983, after which time keyboard sales began to grow at an astonishing rate.

With the adoption of this industry-wide standard, *"any"* device that incorporates MIDI ports into its design can transmit and/or respond to the digital performance and control-related data that conforms to the MIDI 1.0 specification. With this universal fact, you can be assured that the basic functions of any new device will integrate into your existing MIDI system and will work.

Electronic Music Production

Today, MIDI systems are being used by many professional and nonprofessional musicians alike to perform an expanding range of production tasks, including music production, audio-for-video and film postproduction, stage production, etc.

This industry acceptance can, in large part, be attributed to the cost effectiveness, power, and general speed of MIDI production. Once a MIDI instrument or device comes into the production picture, there is often less need (if any at all) to hire outside musicians for a project. This is due to the fact that MIDI's multichannel production environment lets a musician compose, edit, and arrange a piece with a high degree of flexibility, without the need to record and overdub sounds onto multitrack tape (that is, unless you want to).

This affordable potential for future expansion and increased control throughout an integrated production system has spawned the growth of an industry that's also very personal in nature. For the first time in music history, it's possible for an individual to realize cost effectively a full-scale sound production on his or her own time. Because MIDI is a real-time performance medium, it's also possible to listen to and edit a production at every stage of its development, all within the comfort of your own home or personal project studio.

MIDI systems can also be designed to suit your personal production or aesthetic needs. They can be set up to handle specific production tasks (such as the composition of an entire video sound track) with a maximum degree of flexibility and ease, so as to best suit an artist's main instrument and playing style, or even to suit a musician's personal working habits. Each of these advantages is a definite tribute to the power and flexibility that's inherent within modern music production using MIDI.

MIDI in the Home

Currently, a vast number of electronic musical instruments, effects devices, computer systems, and other MIDI-related devices are available on the new and used electronic music market. This diversity lets us select the type of device that best suits our own particular musical taste and production style. With the advent of the *large-scale integrated circuit chip* (which allows complex circuitry to be quickly and easily mass produced), many of the devices that make up an electronic music system are affordable to almost every musician or composer, whether he or she is a working professional, aspiring artist, or beginning hobbyist (Figure 1.3).

MIDI production systems can appear in any number of shapes and sizes and can be designed to match a wide range of production and budget needs. For example, a keyboard instrument (commonly known as a *MIDI workstation*) will often integrate a keyboard, polyphonic synthesizer, percussion sounds, and a built-in sequencer into a single package. Essentially, it's a portable, all-in-one system that lets you quickly and cost effectively create MIDI-style production projects. Additional MIDI instruments can easily be added to this workstation. Simply plug its MIDI out port to the MIDI in port of the new device (or devices), plug the extra audio channels into your mixer, and you're in business.

Figure 1.3
A small home project studio. (Courtesy of Mackie Designs, www.mackie.com)

Other MIDI systems, made up of discrete instruments and devices, are often carefully selected by the artist to generate specific sounds or to serve a particular production function. Although this type of MIDI system isn't very portable, it's often more powerful, because each component combines to create a vast palette of sounds and to handle a wide range of task-specific functions. Such a system might include one or more keyboard synthesizers, synth modules, samplers, drum machines, a computer (with a sequencer and other forms of MIDI and hard disk recording software), effects devices, and audio mixing capabilities.

Systems like these are constantly being installed in the homes of working and aspiring musicians. They can range from those that take up a corner of an artist's bedroom, to larger systems that have been installed in a dedicated project studio. All of these can be functionally designed to handle a multitude of applications, and have the important advantage of letting the artist produce his or her music in a comfortable environment, whenever the creative mood hits. Such production luxuries, which would have literally cost an artist a fortune only a decade ago, are now within the reach of almost every musician.

MIDI in the Studio

MIDI has also dramatically changed the sound, technology, and production habits of the professional recording studio (Figure 1.4). Before MIDI and the concept of the home project studio, the recording studio was one of the

Figure 1.4
A MIDI-equipped recording studio. (Courtesy of Mackie Designs, www.mackie.com)

only production environments that would allow an artist or composer to combine instruments and sound textures into a final recorded product. Often, the process of recording a group in a live setting was (and still is) an expensive and time-consuming process. This is due to the high cost of hiring session musicians and the high hourly rates that are charged for a professional studio—not to mention Murphy's law, which states that you'll always spend more time and money than you thought you ever could capturing the elusive "ideal performance" on tape.

With the advent of MIDI, modular digital multitracks (MDMs), and hard disk recording, much of the music production process can be preplanned and rehearsed, or even totally produced and recorded before stepping into the studio. This out-and-out luxury has reduced the number of hours that are needed for laying tracks onto multitrack tape to a cost-effective minimum. For example, it's now commonplace for groups to record and produce an entire album in their own project studio. Once completed (or nearly completed), the group can either dump the tracks to tape, or simply bring their entire MIDI and recorded audio tracks into the studio and lay the instrument tracks down to tape. In the studio, the tracks can be "sweetened" into a polished state by adding vocals or other instruments. Alternatively, the MIDI tracks may not be recorded onto multitrack tape at all; instead they can be synchronized to tape, allowing the sequencer to act as an extension of the recorded tracks. Whatever the chosen medium, the completed project can be mixed down into a final product in a much more timely fashion than would otherwise be possible.

MIDI in Audio-for-Video and Film

Electronic music has long been an indispensable tool for the scoring and audio postproduction of TV commercials, industrial videos, TV, and full-feature motion picture sound tracks. For productions that are on a tight budget, an entire score can be created in the artist's project studio using MIDI, hard disk tracks, and digital recorders at a mere fraction of what it might otherwise cost to hire the musicians, a studio, and mixdown rooms. Many high-budget projects even make extensive use of MIDI in the preproduction and production phases. Often, producers of high-budget films that use live instruments (often orchestral in nature) will be able to hear a MIDI version of the composer's score before the tracks have been recorded in the studio. Before MIDI, this simply wasn't possible. Once approved, the final MIDI score can be printed and distributed to the musicians before the session.

MIDI in Live Performance

Electronic music production and MIDI are also at home on the stage. In addition to using preprogrammed drum machines and synthesizers on stage, MIDI instruments and effects devices have the added advantage of being able to change their parameters in real time, so that all of the necessary settings for the next song (or section within a song) can be automatically or manually changed using a single program change command.

MIDI's on-stage popularity is primarily due to two factors:

1. Songs can be programmed at home or in the studio, before going onstage.
2. A multitude of instruments, recording systems, and effects devices can be easily controlled from a central location.

The ability to sequence rhythm and background parts in advance, chain them together into a single, controllable sequence (using a jukebox-type sequencing program), and then play them on stage has become an indispensable live-performance tool for many musicians. This technique is currently in use by solo artists who have become one-man bands by adding background sequences (often consisting of a drum machine and a small host of other instruments) to accompany their live vocals and instrument playing. Larger techno-pop groups commonly use extensive on-stage sequencing to drive instruments and effects in addition to those that are played live.

Many live performers make use of MIDI's on-stage real-time control capabilities. This includes the ability to play and have parameter control over a range of instruments from a single performance controller. Different musical and percussion parts can even be played from a single keyboard controller using techniques known as *key and velocity zoning* (whereby different areas or "zones" on the keyboard can be used to communicate to different instruments, depending on what range of keys is being played, or even how hard the keys are being played).

In addition to simply communicating performance data, MIDI can also control a number of device setup parameters in real time. For example, a MIDI controller (such as a master keyboard or MIDI footswitch controller) can be used to change the sound patch and performance parameter settings (i.e., volume, panning, etc.) of various MIDI instruments or effects devices.

MIDI is often directly or indirectly involved in the creation and production of many a live performance. Through the use of sequencers and music notation programs, MIDI can often help a composer input, edit, and audition a musical score. Once finalized, the sequenced score can be printed out as hard copy for distribution to the live, on-stage players.

MIDI and Multimedia

One of the "media" in multimedia is definitely MIDI. It often pops up in places that you might expect and in others that might take you by surprise.

With the advent of General MIDI (a standardized specification that makes it possible for any soundcard or "GM" compatible device to play back a score using the originally intended sounds and program settings), it's possible (and common) for MIDI scores to be integrated into multimedia games, text documents, CD-ROMs, and even web sites. Due to the fact that MIDI is simply a series of performance commands (unlike digital audio, which actually encodes the audio information), the media's data overhead requirements are extremely low. This means that almost no processing power is required to play MIDI, making it the ideal medium for playing real-time music scores while you're actively browsing text, graphics, or other media over the Internet. Truly, when it comes to weaving MIDI into the various media types, the sky (and your imagination) is the creative and technological limit.

2

MIDI 1.0

The Musical Instrument Digital Interface is a digital communications protocol. That is to say, it's a standardized control language and hardware specification that makes it possible for electronic instruments and other device types to communicate performance and control data in real time.

Exploring the Specification

MIDI is a specified data format that must be strictly adhered to by those who design and manufacture MIDI-equipped instruments and devices. Because the format is standardized, you don't have to worry about whether the MIDI output of one device will be understood by the MIDI in port of a device that's made by another manufacturer. As long as the data ports say MIDI, you can be assured that the data (at least the basic performance functions) will be transmitted and understood by all devices within the connected system. In this way, the user need only consider the day-to-day dealings that go hand in hand with using electronic instruments, without having to be concerned with the transmission medium.

The Digital Word

One of the best ways to gain insight into the MIDI specification can be done by comparing MIDI to a spoken language.

Figure 2.1
Meaning is given to the alphabet letters *T*, *I*, and *E* when they're grouped into a word and/or placed into a sentence.

T,I,E = (TIE) =

(alpha-bits)   (word)   (physical equivalent)

As humans, we've adapted our communication skills to best suit our physical selves. Ever since the first grunt, it's been easiest to communicate through the use of our vocal chords, and we've been doing it ever since. Over time, language was developed, which assigned a standardized meaning to a series of sounds (words). Eventually these words were grouped together to convey a more complex means of communication. In order to record the English language, for example, a standard method of notation was developed that assigned 26 symbols to specific sounds (letters of the alphabet) that, when grouped together, would communicate an equivalent spoken word (Figure 2.1). By stringing these words into complete sentences, more complex communication could be conveyed or written down for later review. For instance, the letters *T*, *I*, and *E* don't mean much when used individually. However, when grouped into a word, they refer to a piece of cloth that's worn around the neck as a social convention. When it's placed into a sentence, the word is given a greater clarity of meaning (e.g., "Who's the dude with the bow tie?").

Microprocessors and computers, on the other hand, are digitally based communications devices that obviously lack vocal chords and ears (although even that's changing). However, because they have the unique advantage of being able to process numbers at a very high rate, the obvious language of choice is the reception and transmission of digital data.

Unlike our base 10 system of counting, computers are limited to communicating with a binary system of 0's and 1's (or on and off). Like humans, most computers can group these binary digits (known as *bits*) into larger numeric "words" that represent and communicate specific information and instructions. Just as humans can communicate using simple sentences, a computer can generate and respond to a series of related digital words that are understood by other digital systems or software programs (Figure 2.2).

Figure 2.2
Example of a digitally generated MIDI message.

Status Byte	Data Byte #1	Data Byte #2
(1001 0100)	**(0100 000)**	**(0101 1001)**

The MIDI Message

MIDI digitally communicates musical performance data between devices as a string of MIDI messages. These messages are transmitted through a single MIDI line at a speed of 31.25 kbaud (bits/sec). This data can only travel over a single MIDI line in one direction, from a single source to a destination (Figure 2.3A). To make two-way communication possible, a second MIDI data line must be used to communicate data back to the first device (Figure 2.3B).

MIDI messages are made up of groups of 8-bit words (known as *bytes*), which are transmitted in a serial fashion to convey a series of instructions to one or all MIDI devices within a system.

Only two types of bytes are defined by the MIDI specification: the *status byte* and the *data byte*. A status byte is used as an identifier for instructing the receiving device as to which particular MIDI function and channel is being addressed. Data byte information is used to encode the actual numeric values that are attached to the accompanying status byte. Although

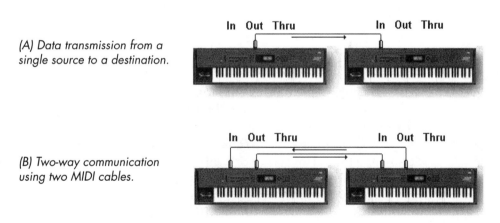

(A) Data transmission from a single source to a destination.

(B) Two-way communication using two MIDI cables.

Figure 2.3   MIDI data can only travel in one direction through a single MIDI cable.

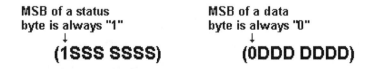

MSB of a status byte is always "1"
↓
(1SSS SSSS)

MSB of a data byte is always "0"
↓
(0DDD DDDD)

Figure 2.4   The most significant bit of a MIDI data byte is used to distinguish between a status byte (where MSB = 1) and a data byte (where MSB = 0).

	Status Byte	**Data Byte 1**	**Data Byte 2**
Description	Status/Channel #	Note #	Attack Velocity
Binary Data	(1001.0100)	(0100.0000)	(0101.1001)
Numeric Value	(Note On/Ch #4)	(64)	(89)

Table 2.1   Status and Data Byte Interpretation

a byte is made up of 8 bits, the *most significant bit* (MSB) (the leftmost binary bit within a digital word) is used solely to identify the byte type. The MSB of a status byte is always 1, while the MSB of a data byte is always 0 (Figure 2.4). For example, a 3-byte MIDI note-on message (which is used to signal the beginning of a MIDI note) in binary form might read as shown in Table 2.1. Thus, a 3-byte note-on message of (10010100) (01000000) (01011001) will transmit instructions that would be read as "Transmitting a note-on message over MIDI channel #4, using keynote #64, with an attack velocity (volume level of a note) of 89."

MIDI Channels

Just as a public speaker might single out and communicate a message to one individual in a crowd, MIDI messages can be directed to a specific device or range of devices in a MIDI system. This is done by imbedding a nibble (4 bits) within the status/channel number byte (Figure 2.5), which makes it possible for performance or control information to be communicated to a specific device or one of the sound generators in a device over its own channel. Because this nibble is 4 bits wide, up to 16 discrete MIDI channels can be transmitted through a single MIDI cable.

Figure 2.5
The least significant nibble of the status/channel number byte is used to encode the channel number.

**Final 4-bit Status Byte
"nibble" is used to encode
the MIDI channel number**
↓
(1SSS CCCC)

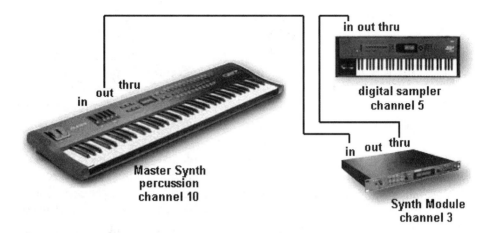

Figure 2.6   MIDI setup showing a set of MIDI channel assignments.

Whenever a MIDI device is instructed to respond to a specific channel number, it will only respond to messages that are transmitted on that channel (i.e., it will ignore channel messages that are transmitted on any other channel). For example, let's assume that we're going to create a short song using a synthesizer that has a built-in sequencer (a device that's capable of recording, editing, and playing back MIDI data) and two other "synths." We could start off by recording a percussion track using channel 10 (which happens to be the same channel that the master synth's percussion sounds are assigned to). Playing back the sequence will transmit the notes and data over channel 10 to the synth's percussion section. Once done, we can set the master synth to transmit notes on channel 3 and then begin recording a melody line into the sequencer. Because the synth module is set to respond to data on channel 3, its generators will sound whenever the master keyboard is played (Figure 2.6). Playing back the sequence will then transmit data to both the master synth and the module over their respective channels. At this point, our song has begun to take shape.

Now, we can set the master synth so that it transmits notes over channel 5, which will then be received by a digital sampler and proceed to play a musical lead line. Now, the song is complete and we can listen to and have control over each instrument (which is set to respond only to its assigned MIDI channel). In short, we have created a true "multichannel" working environment.

MIDI Modes

Electronic instruments often vary in the number of sounds that can be simultaneously produced at a time by their internal sound generating circuitry. For example, certain instruments can only produce one note at a time, while others (known as *polyphonic instruments*) can generate 8, 16, 32, or more notes at once. The latter type lets you play chords and/or more than one musical line on a single instrument.

In addition, various synthesizer types can produce only one characteristic sound patch at a time (e.g., electric piano, synth bass, organ). The word *patch* is a direct reference from earlier analog synthesizers, where *patch chords* where used to connect one sound generator or processor to another. However, it's common for newer instruments to be *multitimbral* in nature (Figure 2.7), meaning that it can generate more than one sound patch (often referred to a "voice") at a time. Thus, it's extremely likely that you already have or will run across electronic instruments that can generate a multitude of voices, each offering its own set of control parameters (such as volume, panning, modulation, etc.) and—best of all—each instrument voice can be assigned to its own MIDI channel!

As a result of the differences between devices, a defined set of guidelines (known as *MIDI reception modes*) has been specified that allows a MIDI instrument to transmit or respond to MIDI channel messages in several ways. For example, one instrument might be programmed to respond to all 16 MIDI channels at one time, while another might be polyphonic in nature, with each voice being programmed to respond to only a single MIDI channel. A listing of these modes are:

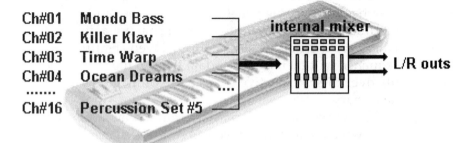

Figure 2.7 Multitimbral instruments are virtual bands-in-a-box in that they can generate more than one polyphonic voice at a time which can be assigned to its own MIDI channel.

Mode 1:	Omni on/poly
Mode 2:	Omni on/mono
Mode 3:	Omni off/poly
Mode 4:	Omni off/mono

Omni on/off refers to how a MIDI instrument will respond to the 16 MIDI channels. When Omni is turned *on*, the MIDI device will respond to all channel messages that are transmitted over all MIDI channels. When Omni is turned *off* the device will only respond to a single MIDI channel or set of assigned channels. *Poly/mono* refers to the sounding of individual notes by a MIDI instrument. In the *poly mode*, an instrument is capable of responding polyphonically to each MIDI channel and is able to produce more than one note at a time. In the *mono mode*, an instrument will respond monophonically to each MIDI channel and will produce only one note at a time. The following list and figures explain the modes in more detail.

- **Mode 1—Omni on/poly:** An instrument will be able to respond polyphonically to performance data that is received on any MIDI channel (Figure 2.8).
- **Mode 2—Omni on/mono:** An instrument will assign any received note events to one monophonic voice, regardless of which MIDI channel it is received on (Figure 2.9). This mode is rarely used.

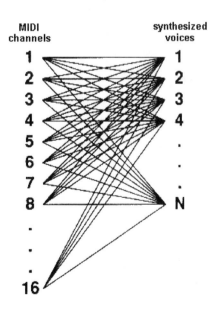

Figure 2.8

Voice/channel assignment example of mode 1 (Omni on/poly).

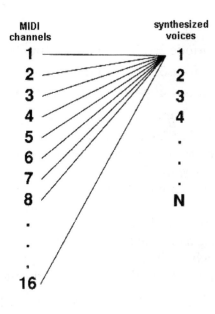

Figure 2.9
Voice/channel assignment example of mode 2 (Omni on/mono).

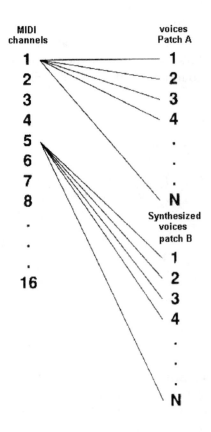

Figure 2.10
Voice/channel assignment example of mode 3 (Omni off/poly).

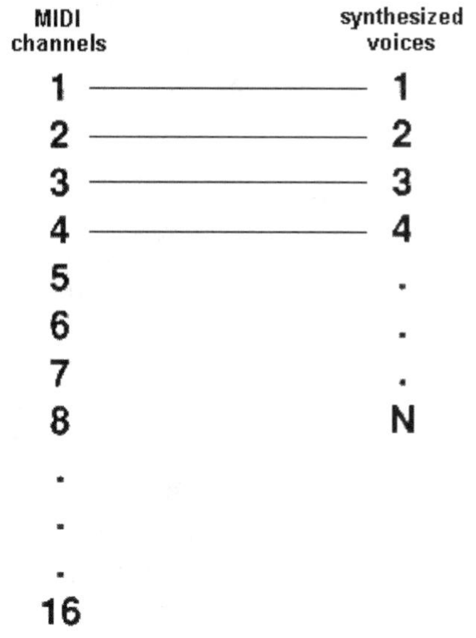

```
        MIDI              synthesized
      channels              voices

        1 ─────────────────── 1
        2 ─────────────────── 2
        3 ─────────────────── 3
        4 ─────────────────── 4
        5                     .
        6                     .
        7                     .
        8                     N

        .

        .

        .
        16
```

Figure 2.11
Voice/channel assignment
example of mode 4 (Omni
off/mono).

- **Mode 3—Omni off/poly:** An instrument (or generated voice) will respond polyphonically to performance data that is transmitted over a single assigned channel (Figure 2.10). As such, this is commonly used by polytimbral instruments.

- **Mode 4—Omni off/mono:** An instrument will respond to performance data that is transmitted over a single assigned channel; however, each voice will only be able to generate one MIDI note at a time (Figure 2.11). A practical example of this mode is often used in MIDI guitar systems, where MIDI data is monophonically transmitted over six consecutive channels (one channel/voice per string). Other electronic instruments can use this mode and allow individual voices to be assigned to any MIDI channel combination.

Channel-Voice Messages

Channel-voice messages are used to transmit real-time performance data throughout a connected MIDI system. They're generated whenever a MIDI instrument's controller is played, selected, or varied by the performer. Examples of such control changes could be the playing of a keyboard, program selection buttons, or movement of modulation or pitch wheels. Each channel-voice message contains a MIDI channel number within its status byte, so that only devices that are assigned to the same channel number will respond to these commands. There are seven channel-voice message types: note-on, note-off, polyphonic-key pressure, channel pressure, program change, control change, and pitch-bend change.

Note-On Messages

A *note-on message* is used to indicate the beginning of a MIDI note. It is generated each time a note is triggered on a keyboard, drum machine, or other MIDI instrument (i.e., by pressing a key, hitting a drum pad, or playing a sequence).

A note-on message consists of three bytes of information (Figure 2.12): a note-on status/MIDI channel number, MIDI pitch number, and attack velocity value. The first byte in the message specifies a note-on event and a MIDI channel (1–16). The second byte is used to specify which of the possible 128 notes (numbered 0–127) will be sounded by an instrument. In general, MIDI note number 60 is assigned to the middle C key of an equally tempered keyboard, while notes 21 to 108 correspond to the 88 keys of an extended keyboard controller.

The final byte is used to indicate the velocity or speed at which the key was pressed (over a value range that varies from 1 to 127). *Velocity* is used to denote the loudness of a sounding note which increases in volume with higher velocity values. Not all instruments are designed to interpret the entire range of velocity values (as with certain drum machines), and others don't respond dynamically at all. Instruments that don't support velocity information will generally transmit an attack velocity value of 64 for every note that's played, regardless of the how soft or hard the keys are actually being pressed. Similarly, instruments that don't respond to velocity messages will interpret all MIDI velocities as having a value of 64.

A note-on message that contains an attack velocity of 0 (zero) is generally equivalent to the transmission of a note-off message. This tells the device to silence a currently sounding note by playing it with a velocity (volume) level of 0.

Status/Ch# (0-15)	Note # (0-127)	Attack Velocity (0-127)
(1001 CCCC)	(0NNN NNNN)	(0VVV VVVV)

Figure 2.12
Byte structure of a MIDI note-on message.

Status/Ch#   Note #   Release Velocity
(0-15)   (0-127)   (0-127)

(1001 CCCC) (0NNN NNNN) (0VVV VVVV)

Figure 2.13
Byte structure of a MIDI
note-off message.

Note-Off Messages

A *note-off message* is used as a command to stop playing a specific MIDI note. Each note-on message will continue to play until a corresponding note-off message for that note has been received. In this way, the bare basics of a musical composition can be encoded as a series of MIDI note-on and note-off events. It should also be pointed out that a note-off message won't cut off a sound; it will merely stop playing it. If the patch being played has a release (or final decay) slope, it will begin that stage upon receiving this message.

As with the note-on message, the note-off structure consists of 3 bytes of information (Figure 2.13): a note-off status/MIDI channel number, MIDI note number, and a release velocity value.

In contrast to the dynamics of attack velocity, the release velocity value (0–127) indicates the velocity or speed at which the key was released. A low value indicates that the key was released very slowly, whereas a high value shows that the key was released quickly. Although not all instruments generate or respond to MIDI's release velocity feature, instruments that are capable of responding to these values can be programmed to vary a note's speed of decay, often reducing the signal's decay time as the release velocity value is increased.

Polyphonic-Key Pressure Messages

Polyphonic-key pressure messages are commonly transmitted by instruments that respond to pressure changes that are applied to the individual keys of a keyboard. Such an instrument can be used to transmit individual pressure messages for each key that's depressed (Figure 2.14).

A polyphonic-key pressure message consists of 3 bytes of information (Figure 2.15): the polyphonic-key pressure status/MIDI channel number, MIDI note number, and pressure value. How a device responds to these messages often varies from manufacturer to manufacturer. However, pressure values are commonly assigned to such performance parameters as

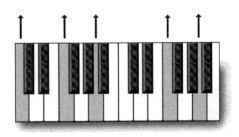

Figure 2.14
Individual polyphonic-key pressure messages are generated when additional pressure is applied to each key that is played.

Status/Ch#   Note #   Pressure Value
(1010 CCCC) (0NNN NNNN) (0VVV VVVV)

Figure 2.15
Byte structure of a MIDI polyphonic-key pressure message.

vibrato, loudness, timbre, and pitch. Although controllers that are capable of producing polyphonic pressure are generally more expensive, it's not unusual for an instrument to respond to these messages.

Channel-Pressure (After Touch) Messages

Channel-pressure messages (often referred to as *after touch*) are commonly transmitted by instruments that only respond to a single, overall pressure, regardless of the number of keys that are being played at one time (Figure 2.16). For example, if six notes are played on a keyboard and additional pressure is applied to only one key, an assigned parameter (such as vibrato, loudness, filter cutoff, and pitch) would be applied to all six notes.

single channel pressure message

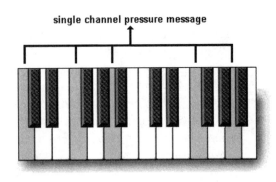

Figure 2.16
Channel-pressure messages simultaneously affect all notes that are being transmitted over a MIDI channel.

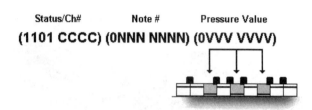

Status/Ch#   Note #   Pressure Value

(1101 CCCC) (0NNN NNNN) (0VVV VVVV)

Figure 2.17
Byte structure of a MIDI
channel-pressure message.

A channel-pressure message consists of 3 bytes of information (Figure 2.17): channel-pressure status/MIDI channel number, MIDI note number, and pressure value.

Program-Change Messages

Program-change messages are used to change a MIDI instrument or device's active program or preset number. A *preset* is a user- or factory-defined number that actively selects a specific sound patch or system setup. Using this message, up to 128 presets can be remotely selected from another device or controller. A program-change message (Figure 2.18) consists of two bytes of information: program-change status/ MIDI channel number (1–16) and a program ID number (0–127).

As an example, a program-change message can be used to switch between the various sound patches of a synthesizer from a remote keyboard instrument (Figure 2.19). It can also be used to select rhythm patterns and/or setups in a drum machine or to call up specific patches on an effects device or any number of controller and system setups that can be recalled as a program preset.

Control-Change Messages

Control-change messages are used to transmit information that relates to real-time control over the performance parameters of a MIDI instrument. Three types of real-time controls can be communicated via control-change messages:

Status/Ch#   Program ID Number

(1100 CCCC) (0PPP PPPP)

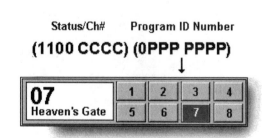

Figure 2.18
Byte structure of a MIDI
program-change message.

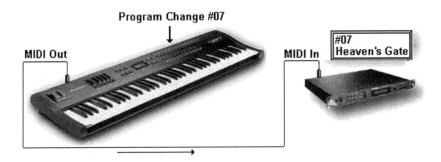

Program Change #07

MIDI Out

MIDI In

#07
Heaven's Gate

Figure 2.19 Program-change messages can be used to change sound patches from a remote controller.

1. **Continuous controllers:** Controllers that relay a full range of variable control settings (often ranging in value 0-127). In certain cases, two controller messages can be combined in tandem to achieve a greater resolution.

2. **Switches:** Controllers that have either an *on* or *off* state, with no intermediate settings.

3. **Data controllers:** Controllers that enter data either through the use of a numerical keypad, or are stepped up and down via data-entry buttons.

A single control-change message or a stream of such messages is transmitted whenever controllers (such as foot switches, foot pedals, pitch-bend wheels, modulation wheels, breath controllers, etc.) are varied in real time. In this way, a wide range of instrument or device parameters can be varied via MIDI, according to the original controller's movements or commands. A control-change message (Figure 2.20) consists of 3 bytes of information: the control-change status/MIDI channel number (1–16), a controller ID number (0–127), and corresponding controller value (0–127).

Status/Ch#	Controller ID #	Controller Value
(1011 NNNN)	**(0CCC CCCC)**	**(0VVV VVVV)**

Figure 2.20
Byte structure of a MIDI
control-change message.

0......64...127

14-BIT CONTROLLER MOST SIGNIFICANT BIT			7 BIT CONTROLLERS (continued)		
Controller *Hex*	*Number* *Decimal*	*Description*	*Controller* *Hex*	*Number* *Decimal*	*Description*
00H	0	Undefined	.	.	.
01H	1	Modulation Controller	4FH	79	Undefined
02H	2	Breath Controller	50H	80	General Purpose Controller #5
03H	3	Undefined	51H	81	General Purpose Controller #6
04H	4	Foot Controller	52H	82	General Purpose Controller #7
05H	5	Portamento Time	53H	83	General Purpose Controller #8
06H	6	Data Entry MSB	54H	84	Undefined
07H	7	Main Volume	.	.	.
08H	8	Balance Controller	.	.	.
09H	9	Undefined	5AH	90	Undefined
0AH	10	Pan Controller	5BH	91	External Effects Depth
0BH	11	Expression Controller	5CH	92	Tremolo Depth
0CH	12	Undefined	5DH	93	Chorus Depth
.	.	.	5EH	94	Celeste (Detune) Depth
.	.	.	5FH	95	Phaser Depth
0FH	15	Undefined	**PARAMETER VALUE**		
10H	16	General Purpose Controller #1	*Controller* *Hex*	*Number* *Decimal*	*Description*
11H	17	General Purpose Controller #2	60H	96	Data Increment
12H	18	General Purpose Controller #3	61H	97	Data Decrement
13H	19	General Purpose Controller #4	**PARAMETER SELECTION**		
14H	20	Undefined	*Controller* *Hex*	*Number* *Decimal*	*Description*
.	.	.	62H	98	Non-Registered Parameter Number LSB
1FH	31	Undefined	63H	99	Non-Registered Parameter Number MSB
14-BIT CONTROLLER LEAST SIGNIFICANT BIT			64H	100	Registered Parameter Number LSB
Controller *Hex*	*Number* *Decimal*	*Description*	65H	101	Registered Parameter Number MSB
20H	32	LSB Value for Controller 0	**UNDEFINED CONTROLLERS**		
21H	33	LSB Value for Controller 1	*Controller* *Hex*	*Number* *Decimal*	*Description*
22H	34	LSB Value for Controller 2	66H	102	Undefined
.	.	.	.	.	.
3EH	62	LSB Value for Controller 30	.	.	.
3FH	63	LSB Value for Controller 31	78H	120	Undefined
7-BIT CONTROLLERS			**RESERVED FOR CHANNEL MODE MESSAGES**		
Controller *Hex*	*Number* *Decimal*	*Description*	*Controller* *Hex*	*Number* *Decimal*	*Description*
40 H	64	Damper Pedal (sustain)	79H	121	Reset All Controllers
41H	65	Portamento On/Off	7AH	122	Local Control On/Off
42H	66	Sostenuto On/Off	7BH	123	All Notes Off
43H	67	Soft Pedal	7CH	124	Omni Mode Off
44H	68	Undefined	7DH	125	Omni Mode On
45H	69	Hold 2 On/Off	7EH	126	Mono Mode On (Poly Mode Off)
46H	70	Undefined	7FH	127	Poly Mode On (Mono Mode Off)
.	.	. *continues*			

Figure 2.21 Listing of controller ID numbers, outlining both the defined format and conventional controller assignments.

Controller ID Numbers

The second byte of the control-change message is used to denote the controller ID number. This number is used to specify which of the device's program or performance parameters are to be addressed.

Although most manufacturers follow a general convention for assigning controller numbers to an associated parameter, they're free to assign them as they wish, provided they follow the defined format as provided by the MIDI specification (Figure 2.21).

Controller Values

The third byte of the control-change message is used to denote the controller's actual data value. This value is used to specify the position, depth, or level that the controller will have on a parameter. In most cases, the value range of a 7-bit continuous controller will fall between 0 (minimum value) and 127 (maximum value) (Figure 2.22). The value range of a switch controller is often 0 (off) and 127 (on) (Figure 2.23A). However, some switching functions can respond to continuous-controller messages by recognizing the values of 0 to 63 as off, and 64 to 127 as on (Figure 2.23B).

The practice of using the value range of 0 to 127 to represent an increasing effect depth or signal level doesn't quite pertain to the control parameters of balance, panning, and expression.

A *balance controller* is used to vary the relative levels between two independent sound sources (Figure 2.24). As with the balance control on a stereo preamplifier, this controller is used to set the relative left/right balance of a stereo signal. The value range of this controller falls between 0 (full left sound source) and 127 (full right sound source), with a value of 64 representing an equally balanced stereo field.

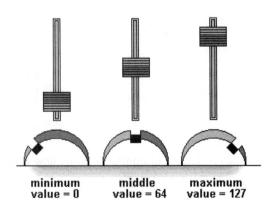

Figure 2.22

Continuous-controller data value ranges.

minimum value = 0   middle value = 64   maximum value = 127

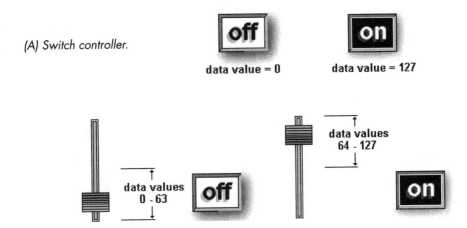

(A) *Switch controller.*

data value = 0

data value = 127

data values
64 - 127

data values
0 - 63

(B) *Some switches respond to a range of continuous-controller messages.*

Figure 2.23 Switch-controller data value ranges.

A *pan controller* is used to position the relative balance of a single sound source between the left and right channels of a stereo sound field (Figure 2.25). The value range of this controller falls between 0 (hard left) and 127 (hard right), with a value of 64 representing a balanced center position.

An *expression controller* is used to accent the existing level settings of a MIDI instrument or device. This control can be used to increase the channel volume level of an instrument, but it can't reduce this level below the programmed volume setting. The value range of this controller falls between 0 (current programmed volume setting) and 127 (full volume accent).

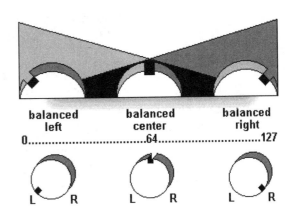

balanced
left

balanced
center

balanced
right

0...64...127

Figure 2.24
Balance-controller data value ranges and corresponding settings.

L   R   L   R   L   R

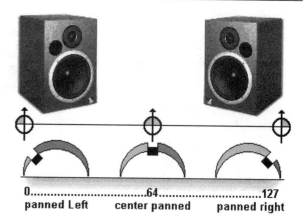

Figure 2.25
Pan-controller data value ranges.

0...64...............................127
panned Left   **center panned**   **panned right**

Controller ID Format
The following controller ID formats have been defined by the MIDI 1.0 specification in order to ensure compatibility between manufacturers with respect to various controller parameters and their messages.

- **14-Bit controllers:** Controller numbers 0 to 31 are commonly reserved for standard continuous-controller messages (such as modulation, breath, main volume, pan, etc.). However, should a resolution be required that's greater than the 128 steps offered by a 7-bit message, it is possible to add an extra 7-bit controller message (using any of the controllers within the 32–63 message number range). Thus, instead of transmitting 2 data bytes over one message, 4 bytes will be transmitted over two messages. This will raise the resolution from 128 to an overall total of 16,384 steps.

- **7-Bit controllers:** Controller numbers 64 to 95 are commonly reserved for switching functions (such as damper pedal, portamento, and sustain), or for those related to the depth of a programmed effect (such as tremolo, chorus, and phase). Because additional messages aren't reserved for these functions their value range is limited to 0 to 127.

- **General-purpose controllers:** These messages aren't assigned to any particular controller function and can be assigned to any device-specific controller at the manufacturer's discretion.

- **Undefined controllers:** Undefined messages aren't currently assigned to any particular controller function. However, they are reserved for future controller parameters.

- **Data-increment controllers:** These messages are used to transmit increment (+) and decrement (–) data to related controls.

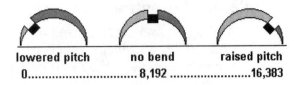

Figure 2.26
Pitch-bend wheel data
value ranges.

lowered pitch   no bend   raised pitch

0.............................. 8,192 16,383

Figure 2.27
Byte structure of a pitch-
bend message.

Status/Ch#   Pitch Bend LSB   Pitch Bend MSB

(1111 NNNN) (0LLL LLLL) (0MMM MMMM)

- **Registered and nonregistered controllers:** Only three types of registered and nonregistered controllers are currently assigned to a controller function: pitch-bend sensitivity, fine tuning, and coarse tuning.

Pitch-bend sensitivity refers to the response sensitivity (in semitones) of a pitch-bend wheel. The 7-bit value range of this controller allows adjustments in increments of up to 1/128 of the overall pitch-bend range. Fine tuning permits controlled adjustments (either up or down) by dividing a semitone into 8,192 possible pitch steps, while course tuning allows for a maximum tuning range of up to 63 semitones above and 64 semitones below the standard tuning pitch.

Pitch-Bend Change

Pitch-bend change messages are transmitted by an instrument whenever its pitch-bend wheel (Figure 2.26) is moved either in the positive (raise pitch) or negative (lower pitch) direction from its central (no-bend) position.

Pitch-bend messages consist of 3 bytes of information (Figure 2.27): a MIDI channel number, least significant bit (LSB) value, and most significant bit (MSB) value. Given that this data is transmitted over 2 data bytes, this message has an overall 14-bit resolution of 16,384 steps.

Channel-Mode Messages

Controller numbers 121 to 127 are reserved for *channel-mode messages*. These include reset-all-controllers, local-control, all-notes-off, and MIDI mode messages.

Reset-All-Controllers Messages

Message number 121 is used to reinitialize all of the controllers (continuous, switch, and incremental) within one or more receiving MIDI instruments or devices to their standard, power-up default state.

Local-Control Messages

The *local-control message* (number 122) is used to disconnect the controller of a MIDI instrument from its own internal voice generators (Figure 2.28). This feature is useful for turning a keyboard instrument into a master controller, by disconnecting an instrument's sound generating circuitry from its own controller (when the instrument's local control is switched off). For example, a synthesizer can be used to control other MIDI devices without having to listen to the synth's own internal sounds. With local control switched off, it's possible to output controller data over a specific MIDI channel to other instruments in the system. By sending performance data to a sequencer, you could easily route MIDI through your computer to other instruments, or back to the controller's own internal voice generators. The latter can be done by simply transmitting data on the same MIDI channel that the synth is set to respond to. In short, the local-control feature splits

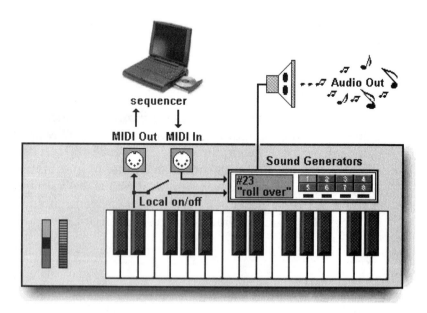

Figure 2.28   The local-control on/off function is used to disconnect a device's controller from its internal sound generators.

your instrument into two parts: a master controller for playing other instruments in the system, and a performance instrument that can be viewed as any other instrument in the MIDI setup. Its effectively makes the system more flexible, while reducing sonic and system conflicts.

A local-control message consists of 2 bytes of information: a MIDI channel number (1–16), and a local-control on/off status byte.

All-Notes-Off Messages

Occasionally, a note-on message will be received by a MIDI instrument and the following note-off message is somehow ignored or not received. This unfortunate event often results in a "stuck note" that will continue to sound until a note-off message is received for that note. As an alternative to endlessly searching for the right note-off key on the right MIDI channel, an all-notes-off "panic message" (message number 123) can often be transmitted, which effectively turns off all 128 notes. Often, a MIDI interface will include a button that will globally transmit this message on all of its ports.

MIDI Mode Messages

Omni Mode Off

Upon the reception of an *Omni mode-off message* (number 124), a MIDI instrument or device will switch modes (or remain in the Omni off mode), so that it responds to individually assigned MIDI channels instead of responding to all MIDI channels at once.

Omni Mode On

Upon receiving an *Omni mode-on message* (number 125), a MIDI instrument or device will switch modes (or remain in the Omni on mode), so that it will respond to all MIDI channel messages, regardless of which channel the messages are being transmitted on.

Mono Mode On

Upon receiving a *mono mode-on message* (number 126), a MIDI instrument will assign individual voices to consecutive MIDI channels, starting from the lowest currently assigned or "base channel." In this way, the instrument can play only one note per MIDI channel, although it is capable of playing more than one monophonic channel at a time.

Poly Mode On

Upon the reception of a *poly mode-on message* (number 127), a MIDI instrument or device will switch modes (or remain in the poly on mode), so that the instrument will respond to MIDI channels polyphonically. This allows the device to play more than one note at a time over a given channel or number of channels.

System Messages

As the name implies, *system messages* are globally transmitted to every MIDI device in the MIDI chain. This is accomplished because MIDI channel numbers aren't addressed within the byte structure of a system message. Thus, any device will respond to these messages, regardless of its MIDI channel assignment. The three system message types are system-common messages, system real-time messages, and system-exclusive messages.

System-Common Messages

System-common messages are used to transmit MIDI time code, song position pointer, song select, tune request, and end-of-exclusive data messages throughout the MIDI system or 16 channels of a specified MIDI port.

MTC Quarter-Frame Messages
MIDI time code (MTC) provides a cost effective and easily implemented way to translate *SMPTE (a standardized synchronization time code)* into an equivalent code that conforms to the MIDI 1.0 spec. It allows time-based codes and commands to be distributed throughout the MIDI chain in a cheap, stable, and easy-to-implement way. *MTC quarter-frame messages* are transmitted and recognized by MIDI devices that can understand and execute MTC commands.

A grouping of eight quarter frames is used to denote a complete time-code address (in hours, minutes, seconds, and frames), allowing the SMPTE address to be updated every two frames. Each quarter-frame message contains 2 bytes. The first is a quarter-frame common header, while the second byte contains a 4-bit nibble that represents the message number (0–7). A final nibble is used to encode the time field (in hours, minutes, seconds, or frames). More in-depth coverage of MIDI time code can be found in Chapter 10.

Song Position Pointer Messages
As with MIDI time code, *song position pointer (SPP)* lets you synchronize a sequencer, tape recorder, or drum machine to an external source from any measure position within a song. The *SPP message* is used to "reference" a location point in a MIDI sequence (in measures) to a matching location within an external device. This message provides a timing reference that increments once for every six MIDI clock messages (with respect to the beginning of a composition).

Unlike MTC (which provides the system with a universal address location point), SPP's timing reference can change with tempo variations, often requiring that a special tempo map be calculated in order to maintain synchronization. Because of this fact, SPP is used far less often than MIDI time code.

SPP messages are generally transmitted while the MIDI sequence is stopped, allowing MIDI devices equipped with SPP to chase (in a fast-forward motion) through the song, and lock to the external source once relative sync is achieved. More in-depth coverage of the SPP can be found in Chapter 10.

Song Select Messages

Song-select messages are used to request a specific song from a drum machine or sequencer (as identified by its song ID number). Once selected, the song will thereafter respond to MIDI start, stop, and continue messages.

Tune Request Messages

The *tune request message* is used to request that a MIDI instrument initiate its internal tuning routine (if so equipped).

End-of-Exclusive Messages

The transmission of an *end-of-exclusive (EOX) message* is used to indicate the end of a system-exclusive message. In-depth coverage of system-exclusive messages will be discussed later in this chapter and in Chapter 4.

System Real-Time Messages

Single-byte *system real-time messages* provide the precise timing element required to synchronize all of the MIDI devices in a connected system. To avoid timing delays, the MIDI specification allows system real-time messages to be inserted at any point in the data stream, even between other MIDI messages (Figure 2.29).

Timing-Clock Messages

The MIDI *timing-clock message* is transmitted within the MIDI data stream at various resolution rates. It is used to synchronize the internal timing clocks of each MIDI device within the system and is transmitted in both the start and stop modes at the currently defined tempo rate.

In the early days of MIDI, these rates (which are measured in pulses per quarter note, ppq) ranged from 24 to 128 ppq. However, continued advances in technology have brought these rates up to 240, 480, or even 960 ppq.

Figure 2.29 System real-time messages can be inserted within the byte stream of other MIDI messages.

Start Messages

Upon receipt of a timing-clock message, the MIDI *start command* instructs all connected MIDI devices to begin playing from their internal sequences initial start point. Should a program be in midsequence, the start command will reposition the sequence back to its beginning, at which point it will begin to play.

Stop Messages

Upon receipt of a MIDI *stop command,* all devices within the system will stop playing at their current position point.

Continue Messages

After receiving a MIDI stop command, a MIDI *continue message* will instruct all connected devices to resume playing their internal sequences from the precise point at which it was stopped.

Active-Sensing Messages

When in the stop mode, an optional *active-sensing message* can be transmitted throughout the MIDI data stream every 300 milliseconds. This instructs devices that can recognize this message that they're still connected to an active MIDI data stream.

System-Reset Messages

A *system-reset message* is manually transmitted in order to reset a MIDI device or instrument back to its initial power-up default settings (commonly mode 1, local control on, and all notes off).

System-Exclusive Messages

The *system-exclusive (SysEx) message* lets MIDI manufacturers, programmers, and designers communicate customized MIDI messages between MIDI devices. These messages give manufacturers, programmers, and designers the freedom to communicate any device-specific data of an unrestricted length as they see fit. SysEx data is commonly used for the bulk transmission and reception of program/patch data, sample data, and real-time control over a device's parameters.

The transmission format of a SysEx message (Figure 2.30) as defined by the MIDI standard includes a SysEx status header, manufacturer's ID number, any number of SysEx data bytes, and an EOX byte. Upon receiving a SysEx message, the identification number is read by a MIDI device to determine whether or not the following messages are relevant. This is easily accomplished, because a unique 1- or 3-byte ID number is assigned to each registered MIDI manufacturer. If this number doesn't match the receiving MIDI device, the ensuing data bytes will be ignored. Once a valid stream of SysEx data is transmitted, a final EOX message is sent, after which the device will again begin responding to incoming MIDI performance messages. A detailed practical explanation of the many uses (and wonders) of SysEx can be found in the synthesizer section of Chapter 4, as well as in the patch editor section of Chapter 6. I definitely recommend that you check these out, because SysEx is one of the most cost-effective and powerful tools that an electronic musician can have. It's definitely well worth the reading!

Universal Nonreal-Time System Exclusive

Universal nonreal-time SysEx data is a protocol that's used to communicate control and nonreal-time performance data. It's currently used to intelligently communicate a data-handshaking protocol, which informs a device

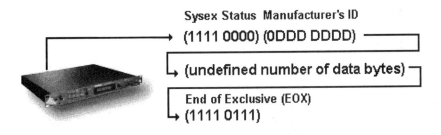

Sysex Status   Manufacturer's ID
(1111 0000) (0DDD DDDD)

(undefined number of data bytes)

End of Exclusive (EOX)
(1111 0111)

Figure 2.30   System-exclusive data (one ID byte format).

that a specific event is about to occur, or that specific data is about to be requested. It is also used to transmit and receive universal sample-dump data, or to transmit MIDI time-code cueing messages. A universal nonreal-time SysEx message consists of 4 or 5 bytes that includes two sub-ID data bytes that identify which nonreal-time parameter is to be addressed. It is then followed by a stream of pertinent SysEx data.

Universal Real-Time System Exclusive

Currently, two universal real-time SysEx messages are defined. Both of them relate to the MTC synchronization code (which is discussed in detail in Chapter 10). These include full message data (relating to a SMPTE address) and user-bit data.

Running Status

Within the MIDI 1.0 specification, special provisions have been made to reduce the need for conveying redundant MIDI data. This mode, known as *running status*, allows a series of consecutive MIDI messages that have the same status byte type to be communicated without repeating the same status byte each time a MIDI message is sent. For example, we know that a standard MIDI message is made up of both a status byte and one or more data bytes. When using running status, a series of pitch-bend messages that have been generated by a controller would transmit an initial status and data byte message, followed only by a series of related data (pitch-bend level) bytes, without the need for including redundant status bytes. The same could be said for note-on, note-off, or any other status message type.

Although the transmission of running-status messages is optional, all MIDI devices must be able to identify and respond to this data transmission mode.

3

THE HARDWARE

In addition to the large number of electronic MIDI instruments that are on the market, various MIDI hardware systems also exist for the purpose of interfacing, distributing, processing, and diagnosing MIDI data. These supporting systems are used to integrate all of the individual tools and toys into a working environment that's designed to be cohesive and easy to use.

System Interconnection

As a data transmission medium, MIDI is relatively unique in the world of sound production in that it allows 16 channels of performance, controller, and timing data to be transmitted in one direction over a single MIDI cable. Using this method, it's possible for MIDI messages to be communicated from a specific source (such as a keyboard or MIDI sequencer) to any number of devices within a network over a single MIDI data chain. In addition, MIDI is flexible enough that more than one data line can be used to interconnect devices in a wide range of possible system configurations. Such a multiple data line system can easily transmit MIDI data over 32, 48, 128, or more discrete MIDI channels!

The MIDI Cable

A MIDI cable (Figure 3.1) consists of a shielded, twisted pair of conductor wires that has a male 5-pin DIN plug located at each of its ends. The MIDI specification uses only 3 of the 5 pins, with pins 4 and 5 being used

Figure 3.1 Wiring diagram and picture of a MIDI cable.

as conductors for MIDI data, while pin 2 is used to connect the cable's shield to equipment ground. Pins 1 and 3 are currently not in use, but are reserved for possible changes in future MIDI applications. Twisted cable and metal shield groundings are used to reduce outside interference, such as radio-frequency interference (RFI) or electrostatic interference, which can serve to distort or disrupt MIDI message transmissions.

MIDI cables come prefabricated in lengths of 2, 6, 10, 20, and 50 feet, and can commonly be obtained from music stores that specialize in MIDI equipment. To reduce signal degradations and external interference that tends to occur over extended cable runs, 50 feet is the maximum length specified by the MIDI specification. (As an insider tip, I found that Radio Shack is a great source for picking up 6-foot MIDI cables at a fraction of what you'd spend at a music store. The only difference is that they call them 5-pin DIN cables, which is technically correct. If you walk in asking for a MIDI cable, most stores won't know what you're talking about.)

MIDI Ports

MIDI is distributed from device to device using three types of MIDI ports: MIDI in, MIDI out, and MIDI thru (Figure 3.2). These three ports use 5-pin DIN jacks to provide interconnections between MIDI devices within a connected network: The hardware designs for these ports (as strictly defined by MIDI 1.0 Spec.) are optically isolated to eliminate any possible ground loops that might occur when connecting numerous devices together.

MIDI In Port

The *MIDI in port* receives messages from an external source and communicates this performance, control, and/or timing data to the device's internal microprocessor. More than one MIDI in port can be designed into a system to provide for MIDI merging functions or for devices that can support more than 16 channels (such as a MIDI interface). Other devices (such as a controller) might not have a MIDI in port at all.

(A) Physical port layout.

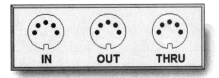

(B) Signal path of MIDI in, out, and thru ports.

Figure 3.2
MIDI in, out, and thru ports.

MIDI Out Port

The *MIDI out port* is used to transmit MIDI performance or control messages from one device to another MIDI instrument or device. More than one MIDI out port can be designed into a system, which has the advantage of giving a system more than 16 channels or allowing the user to filter MIDI data on one port, while not selectively restricting the data flow on another port.

MIDI Thru Port

The *MIDI thru port* retransmits an exact copy of the data that's being received at the MIDI in port. This port is important, because it allows data to pass through an instrument or device to the next instrument or device that follows within the MIDI data chain. Keep in mind that this port is used to relay an exact copy of the MIDI in data stream, and is not merged with data being transmitted at the MIDI out port.

MIDI Echo

Certain MIDI devices don't include a MIDI thru port. Such devices, however, may give you the option of switching the MIDI out between being an actual MIDI out port and a *MIDI echo port* (Figure 3.3). As with the MIDI thru port, a MIDI echo option can be used to retransmit an exact copy of any information that's received at the MIDI in port and route this data to the MIDI out/echo port. Unlike a dedicated MIDI out port, the MIDI echo function

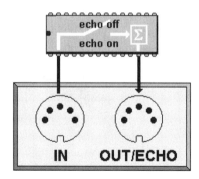

Figure 3.3
MIDI echo configuration.

can often be selected to merge incoming data with performance data that's being generated by the device itself. In this way, more than one controller can be placed in a MIDI system at one time. Note, however, that although performance and timing data can be echoed to a MIDI out port, not all devices can echo SysEx data.

Typical Configurations

Although there are many different types of MIDI setups and lots of equipment types that can make up such a system, a few conventions are followed that make it easy for MIDI devices to be connected. These common configurations allow MIDI data to be distributed in the most efficient manner possible.

As a primary rule, there are only two valid methods of connecting one MIDI device to another (Figure 3.4):

1. Connecting the MIDI out port (or MIDI out/echo port) of one device to the MIDI in port of another device.
2. Connecting the MIDI thru port of one device to the MIDI in port of another device.

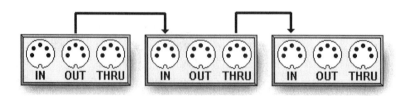

Figure 3.4   The two valid means of connecting one MIDI device to another.

The Daisy Chain

One of the simplest and most common ways to distribute data throughout a MIDI system is the *daisy chain*. This method relays MIDI from one device to the next by retransmitting data that's received at a device's MIDI in port out to another device via its MIDI thru (or MIDI echo) port. In this way, MIDI data can be chained from one device to the next. For example, a typical MIDI daisy chain (Figure 3.5A) will flow from the MIDI out port of a source device (such as a controller, sequencer, etc.) to the MIDI in port of the second device. The MIDI data being received by the second device is also routed through to its MIDI thru port, which is plugged into a third device's MIDI in port. The data received by the third device is then relayed to its MIDI thru port and into a fourth device, and so on until the final device in the chain is reached.

A computer can also easily be designated as the master source in a daisy chain, so that a sequencing program can be used to control the entire playback, channelizing, and signal processing functions of a system. In Figure 3.5B, the

(A) Typical daisy chain hookup.

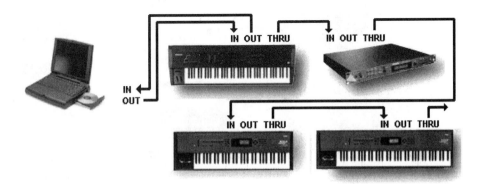

(B) Example of how a computer can be connected into a daisy chain.

Figure 3.5 Example of a connected MIDI system using a daisy chain.

MIDI out port of a master MIDI keyboard controller is routed to the computer. The computer's MIDI out port is then routed back into the controller's MIDI in port, where the data begins its relay throughout the system in a single-line daisy chain fashion.

Because the MIDI thru port relays an exact duplicate of the data that's presented at its MIDI in port, the signal can be traced back through each device to a single master device. In most cases, this source will transmit data over numerous MIDI channels, which are, in turn, individually responded to by devices in the chain that have been assigned to these channels.

The Star Network

Another popular way to integrate devices into a MIDI system is to interconnect them using a *star network* (Figure 3.6). The typical star system strategy allows a master controller to communicate with a computer and chain of MIDI instruments or devices over the individually addressable MIDI ports that are available on newer multiport MIDI interfaces.

In larger, more complex MIDI systems, a star network offers several advantages over a single-line daisy chain network. One of the most important is the ability to independently address the channelizing or data processing needs of each MIDI signal line. For example, port A in a star network might include a device that wants to respond only to channels 1 to 8 (thus requiring that channels 9 to 16 be ignored). On the other hand, devices on port B are able to respond to all 16 channels, but shouldn't receive program change messages. The device that's connected to port C need only respond to channel 1 and prefers not to receive data on any other channel. By using a star network, data processing hardware and/or software can be used to configure the basic channel and processing functions

Figure 3.6
Example of a star network using a multiport MIDI interface.

of each port independently to accommodate the needs of each isolated data stream.

Another bonus offered by a star network is the ability to patch MIDI data between devices that exist on different network ports. This function, which is available on most MIDI patchbays (devices that can selectively route MIDI data paths) and multiport computer interfaces, can often be software or hardware programmed to allow the system to be reconfigured at any time to match the current production needs.

MIDI and the Personal Computer

Besides the coveted place of honor in which most electronic musician's hold their instruments, the most important device in a MIDI system is undoubtedly the personal computer. Through the use of software programs and peripheral hardware, the personal computer (PC) is often used to control, process, and distribute information relating to music performance and production from a centralized, integrated control position.

The personal computer is a high-speed digital processing engine that can perform a wide range of work-, production-, or fun-related tasks. A PC is generally assembled from such off-the-shelf building block components as a power supply, motherboard/central processing unit (CPU), random access memory (RAM), floppy and hard disk(s), and a CD-ROM drive.

Most computers are designed with several hardware slots that let you add expansion cards to the system, so as to perform specific tasks that can't be handled by the computer's own hardware. Using these slots and Universal Serial Bus (USB) ports, devices for connecting your computer to the world of digital audio, MIDI, video, the Internet, scanning, networking, etc., can be easily integrated with the CPU using a 16-, 32-, or 64-bit communications structure.

Software, that all-important computer-to-human interface, gives us countless options for performing all sorts of tasks such as music sequencing, data routing, digital audio editing and playback, signal processing, and music printing. Given the countless software and shareware options that are available for handling and processing most types of data, we're faced with the luxury of being able to choose the type of interface that best suits our personal working style and habits.

When you get right down to it, the fact that a computer can be individually configured to best suit the individual's own personal needs has been one of the driving forces behind the digital age. It has turned the computer into a powerful "digital chameleon" that can change its form and function to fit the task at hand.

For the most part, three personal computer types are used in modern-day MIDI production: the Macintosh, the IBM-compatible, and (to a lesser extent) the Atari. Each brings its own particular set of advantages and disadvantages to personal computing, with the primary differences being personal preference, cost, ease of operation, and hardware/software availability options.

The Mac

One computer type that is widely accepted by music professionals is the Macintosh family of computers from Apple Computer, Inc. One of the major reasons for the initial success of the "Mac" is its graphic user interface, which uses graphic icons and mouse-related commands to offer a friendly environment that lets you move, expand, tile, or stack windowed applications on the system's monitor.

In addition to its ease of use, these powerful computers offer the distinct advantage of allowing hardware to be added to a system with a minimum amount of hardware setup and reconfiguration. Due to the rigid hardware design constraints that are required by the Mac's operating system (OS), many (but not all) of the potential bugs and setup problems that are associated with most PCs are minimized.

The PC

Due to its cost effectiveness, the Windows OS, the large amount of software available for it, and its sheer numbers in both the home and business com-

Figure 3.7
Example of an IBM-compatible PC. (Courtesy of IBM Corporation, www.ibm.com)

munity, the IBM-compatible personal computer (Figure 3.7) clearly dominates the marketplace (musical or otherwise).

Unlike the Mac, which is made by one manufacturer (with a few clone exceptions), the PC's general specifications were licensed out to the industry at large. Because of this fact, countless manufactures make PCs that can be factory assembled or user assembled and upgraded (with a little basic knowledge and a lot of common sense) using standard, off-the-shelf components.

The PC's operating system of choice is the ever-present Windows OS from Microsoft. As with the Mac's OS, Windows is a graphic-based, multitasking environment that can run several programs and task-based applications at a time. Using the newer generation of Pentium processors that can run at speeds in excess of 300 MHz, Windows 95, 98, and later OS platforms can perform complex MIDI digital audio and visual functions with amazing speed, using a full 32-bit processing architecture.

The Atari

Although not as common as they used to be in the United States, the Atari line of personal computers is still alive and well in Europe and around the world. These cost-effective computers offer an easy-to-use, multitasking graphic environment that has instilled a strong sense of devotion among musicians and other media enthusiasts. One of the reasons for this popularity was (and still is) Atari's dedication to cost-effective media production.

Unlike the Macintosh and most IBM-compatible computers, the Atari's design (most notably the ST and Mega line) has MIDI in and out ports designed directly into the computer's housing as standard equipment.

Portability

All of the computer types just discussed are available in configurations that can be easily taken on the road. Just like cellular phones, the laptop has become a common sight in today's business and media production scene. These small, lightweight powerhouses often contain the same CPUs, graphics capabilities, hard disk space, and CD-ROM drives as the most powerful workstations. With the development of multiport MIDI interfaces, high-quality PCMCIA (a special hardware interface often used by laptops) soundcards and modems, the laptop has turned into an on-the-road, battery-operated computer that's as powerful as any of the big boys.

The MIDI Interface

Although computers and electronic instruments both communicate using the digital language of 1's and 0's, computers simply can't understand the language of MIDI without the use of a device that translates the serial messages into a data structure that computers can comprehend. Such a device is know as a *MIDI interface* (Figure 3.8).

A wide range of MIDI interfaces currently exist that can be used with most computer systems and OS platforms. Some interface types that you can expect to find are external passive, soundcard, PC hardware card, synth interface, and external multiport MIDI interfaces.

A passive interface (Figure 3.9) is a simple external device that plugs into a computer's serial or parallel port. Its sole purpose in life is to give you a MIDI in and one or two MIDI out ports, period! Generally, no battery or power supply is needed because this interface type often gets its power from the computer itself. Their distinct advantage is that they are simple, reliable, and cheap.

When talking about MIDI interfaces, who could pass up the one interface that surpasses all others in numbers (by the millions)? I'm referring,

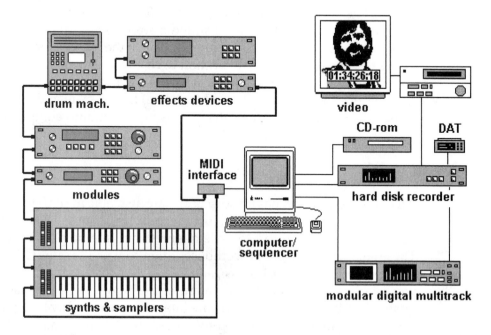

Figure 3.8 Example of a system using a computer/instrument MIDI interface connection.

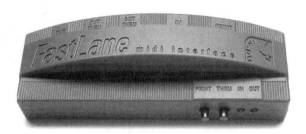

Figure 3.9
MOTU's FastLane passive
interface. (Courtesy of
Mark of the Unicorn, Inc.,
www.motu.com)

of course, to all of those SoundBlaster PC soundcards and their look-alikes
that have a 16-channel MIDI interface built right into them.

Almost every soundcard, no matter how cheap, will have a 15-pin connector that can be used as a game joystick port or for plugging in a MIDI port
cable adapter. These inexpensive adapter cables can then be used to interface the PC to electronic instruments over 16 or 32 channels (depending on
the soundcard's capabilities).

Other interface card types for the PC have been designed to deal ex-
clusively with MIDI data and don't have any digital audio capabilities.
Most MIDI hardware cards for the PC (Figure 3.10) offer such added features over their soundcard counterparts as multiport capabilities and facilities for synchronizing external audio and video recorders to a MIDI
sequencer.

Certain synth keyboards and keyboard modules have MIDI interface
capabilities built right into them. This comes in the form of an interface that
connects the serial port of your computer to the instrument. By installing the

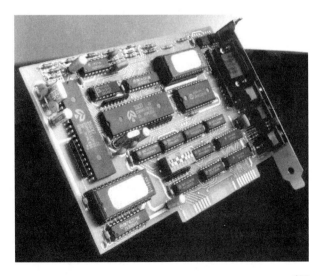

Figure 3.10
Opcode MQS-32M MIDI
interface card for the PC.
(Courtesy of Opcode Systems
Inc., www.opcode.com)

Figure 3.11
The Studio 128X multiport
MIDI interface/synchronizer.
(Courtesy of Opcode Systems
Inc., www.opcode.com)

factory-supplied drivers (or those that you picked up from their web site) into your computer, the instrument's sound generating capabilities and its MIDI ports can be directly accessed by the computer. For the beginner, this means that you don't need to buy a MIDI interface. You can simply connect the instrument's interface to your computer's serial port and then connect your other MIDI instruments to the synth in a simple daisy chain fashion. For those who already have a 16-channel interface, an instrument interface can be used as an additional port, giving you access to 32 or more MIDI channels.

The interface of choice for most professional electronic musicians is the external multiport MIDI interface (Figure 3.11). These rack-mountable devices often plug into the computer's serial or parallel port to provide four, eight, or more independent MIDI in and out ports, which can easily distribute MIDI data through separate lines over a star network.

In addition to multiport capabilities, these software-controlled systems can be used as a patchbay for routing MIDI data from an instrument or device to others that exist on another signal path. Special MIDI processing functions are also commonly found on most multiport interfaces. Examples of these functions include the ability to merge several MIDI inputs into a single data stream, filter out specific MIDI message types (used to block out unwanted commands that might adversely change an instrument's sound or performance), and rechannelize data that's being transmitted on one MIDI channel to another channel that can be recognized by an instrument or device.

Another important function that's handled by most multiport interfaces is synchronization. Synchronization (sync for short) allows other, external devices (such as an analog or digital tape recorder, videocassette recorder, or other types of recorded media) to be played back using the same timing reference, so that MIDI and other performance-related events occur at the same time. In short, synchronization is used to make all of the separate devices within a production system act as a single entity, allowing events on all devices to occur at the same relative time.

An interface that includes sync features will often read and write SMPTE time code, convert SMPTE to MIDI time code (MTC) and allow recorded time code signals to be cleaned up when copying code from one analog device to another. Other interface types might even be able to read the specialized time code formats that are used by modular digital multi-track (MDM) machines. This option let's you synchronize ADAT or other MDM machine types to your MIDI sequencer and instruments without needing to buy additional sync hardware. Further reading on the various types of synchronization can be found in Chapter 10.

MIDI Distribution and Processing

As with all human interface situations, we're always most comfortable when our working space is familiar, straightforward, and stable. This also holds true in MIDI production. Electronic musicians are usually happiest when all the instruments are assigned to their proper channels and all the favorite sound patches, controls, and system setups can be easily accessed without having to think much about it.

Within larger MIDI production systems, however, system setups may need to be changed from time to time and special occasions may arise that require you to alter your system's overall setup (either temporarily or into a new setup pattern, as often happens when a new device or toy is added to the system). Whenever changes need to be made, cables may need to be restrung, data might need to be repatched from one device to another, or the MIDI streams themselves may need to be processed (so as to filter out certain messages that might wreak havoc at just the wrong moment).

The following subsections discuss a series of functions that deal with the day-to-day realities of shuffling MIDI data throughout a MIDI system. Keep in mind that under certain circumstances, special hardware boxes might be required to handle the job. Sometimes you might be able to make these devices yourself with a trip to Radio Shack, a soldering iron, and some commonsense skills; at other times you might need to buy a new piece of equipment. However, you might be surprised to learn that it's not uncommon for your system to have one or more of these processing capabilities built right into your interface. This is especially true for newer generation multiport interfaces that can handle extensive patching, processing, and diagnostic functions. The rule of thumb is to check the manual for the interface to see if you already have the tools you need for the job. Examples of such processing functions include merging, patching, data processing, and diagnostics.

Merging MIDI Data

Whenever data from two or more separate MIDI lines are combined into a single data stream, a *MIDI merger* must be used. This needs to be done because it's not possible to simply splice two MIDI data lines into one input since the data wouldn't be properly synchronized and the resulting collisions would be irretrievably intertwined and not recognizable as valid MIDI data. Therefore, the MIDI merger acts as a traffic manager, allowing incoming data to be interleaved into a continuous MIDI stream that contains valid MIDI messages from each MIDI source.

MIDI mergers (which can be a dedicated device, MIDI patchbay, or multiport interface) allow two or more devices to act jointly as source controllers for other instruments, sequencers, or devices in a MIDI system. For example, in Figure 3.12, the MIDI out data from both a keyboard controller and a computer/sequencer could be used to control an on-stage MIDI system. Using this example, a keyboardist could do live improvisation with a sequenced track while drinking a Blue Hawaiian Fizz in the Tiki Lounge.

Patching MIDI Data

Whenever situations come up that require you to change your MIDI system's signal routing paths, it's generally no big deal to plug and unplug a few MIDI cables, if your system is small. However, if the system includes three or more separate MIDI daisy chains, if your cabling setup resembles

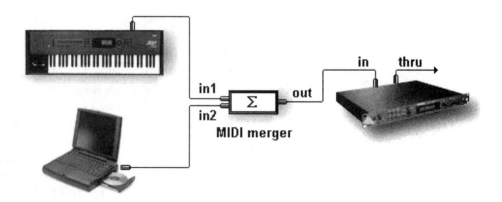

Figure 3.12 MIDI data can be combined into a single data stream using a MIDI merger.

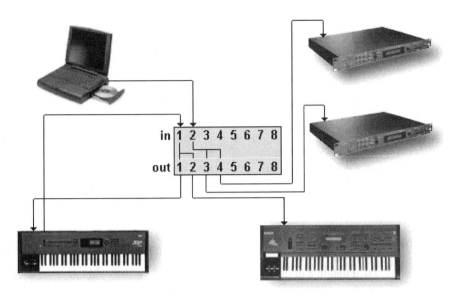

Figure 3.13 Example of how MIDI data can be distributed using a MIDI patchbay.

spaghetti, or if the cables are hard to get to, making some simple changes could lead to a headache (particularly if you have to make them often). It is for these reasons that the *MIDI patchbay* was born.

The function of a MIDI patchbay is to route one or more MIDI data sources to one or more devices throughout a connected system (Figure 3.13). MIDI patchbays come in a wide range of port configurations and operational styles. For example, certain systems are completely manual, requiring that front panel controls be switched in order to change the system's layout configuration. Other types use a microprocessor to route data through the various MIDI paths of a star network. Most multiport MIDI interfaces that are found in project and modern-day production studios offer a patchbay feature that ranges from layouts for setting the system's overall parameters to options that offer extensive patching and data processing options for each port/MIDI data path.

Often modern-day patchbays and multiport interfaces are capable of the following functions:

- Merging incoming data from several inputs to a single MIDI out port
- Distributing MIDI data from a single source to one, several, or all output ports on the network
- Routing data from any controller source to any port/device destination without the need to repatch cables manually

51

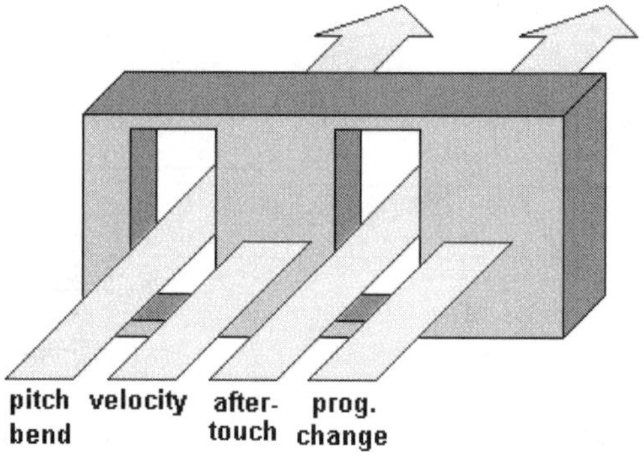

Figure 3.14
A MIDI filter is used to block the transmission of specific MIDI messages.

pitch   velocity   after-   prog.
bend   touch   change

Filtering MIDI Data

Through the use of dedicated devices or a suitably equipped multiport MIDI interface, specific messages or a range of messages within a data stream can be either recognized or ignored. These processor-based systems can be manually or software programmed to block the transmission of specific MIDI messages (Figure 3.14), such as velocity on/off, program-change on/off, modulation on/off, and SysEx on/off.

Whenever an intelligent multiport interface is used, messages that are transmitted over a single MIDI data line and/or a single MIDI channel can be filtered. In this way, only a single device, chain of devices in a data line, or specific instrument voice can be affected.

Mapping MIDI Data

MIDI mapping is a process in which the value of a data byte or range of data bytes can be reassigned to another value or range of values. This function lets you change one or more of the parameters in an existing MIDI data stream to an alternate value or set of values (Figure 3.15). Mapping can be applied to a MIDI data stream to reassign channel numbers, transpose notes, change controller numbers and values, and limit controller values to a specific range.

Figure 3.15
MIDI mapping can be used to change data within the MIDI message stream.

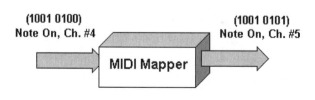

(1001 0100)
Note On, Ch. #4

MIDI Mapper

(1001 0101)
Note On, Ch. #5

As with MIDI filtering, whenever an intelligent multiport interface is used, messages that are transmitted over a single MIDI data line and/or a single MIDI channel can be mapped. In this way, only a single device, chain of devices in a data line, or specific instrument voice will be affected.

Processing MIDI Data

Just as signal processors can be used in an audio chain to create an effect by changing or augmenting an existing sound, a *MIDI processor* can be used to alter MIDI messages within the system. MIDI processing (which can be performed from software using a MIDI sequencer, other MIDI program, or hardware device such as a multiport MIDI interface) can be used to perform such algorithmic functions on one or more MIDI signals as:

- MIDI channel and continuous-controller reassignments
- Programmable MIDI delays
- MIDI message filtering
- Velocity scaling and limiting
- Chromatic transposition
- Note, velocity, and pitch-wheel inversions

MIDI Diagnostic Tools

Although software-based tools exist for analyzing MIDI data from a computer, a number of hardware tools can also be used to detect and diagnose messages as they travel over a single MIDI line or throughout a MIDI production studio or live stage setting. Such tools can be used for detecting the presence of MIDI data or for troubleshooting specific MIDI message types.

Probably the simplest and most practical MIDI diagnostic tool is an LED *(light-emitting diode)* that indicates the presence of incoming or outgoing MIDI data. Such practical indicators have been designed into the front panel of most multiport MIDI interfaces (letting you easily see MIDI activity on all in and out ports). However, these MIDI-in activity LEDs or LCD indicators have also been designed into almost every electronic instrument, allowing you to see if they're receiving data on their assigned channels.

In certain situations, you might not be able to check visually for the presence of MIDI. This occurs if (1) the instrument doesn't have a data indicator or (2) activity indicators on an instrument won't light up if it is receiving data on a MIDI channel to which it's not assigned. In either case, you

Figure 3.16
Diagram detailing how to make your own MIDI activity light.

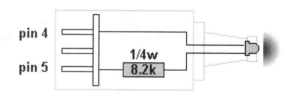

Figure 3.17
The MA36 MIDI analyzer.
(Courtesy of Studiomaster Inc.,
www.studiomaster.com)

could look at the LEDs on your multiport interface or you could buy or build a simple MIDI activity light that will let you know instantly if data activity is happening on that MIDI line.

An example of such a handy tool is the "Ensoniq Spider." This fun, plastic spider has two LEDs for eyes that light up whenever MIDI data is present. The other option would be to make your own by going down to Radio Shack and picking up a single, male 5-pin DIN plug, an 8.2-kohm 1/4-watt resistor, and an LED of your color choice and following the simple diagram shown in Figure 3.16.

Another diagnostic tool that can more thoroughly indicate the presence of specific MIDI message types is the MA36 MIDI analyzer from Studiomaster, Inc. (Figure 3.17). This handheld tool can be used to indicate which MIDI channel(s) are being transmitted over a line. It can also display the type of MIDI messages that are being transmitted or received.

4

ELECTRONIC MUSICAL INSTRUMENTS

Since their inception in the early 1980s, MIDI-based electronic instruments have played a central and important role in the development of music technology and production. These devices (which fall into almost every instrument category), along with the advent of cost-effective analog and digital audio recording systems, have probably been the two most important technological advances to shape the industry into what it is today. In fact, the combination of these technologies has made the personal project studio into one of the most important driving forces behind modern-day music.

Inside the Toys

Although electronic instruments often differ from one another in looks, form, and function, they almost always share a standard set of basic building block components (Figure 4.1), including the following:

- **Central processing units (CPU):** One or more dedicated computers (often in the form of a specially manufactured microprocessor chip) that contain all of the necessary brains to control the hardware, voice data, and sound generating capabilities of the entire instrument or device.

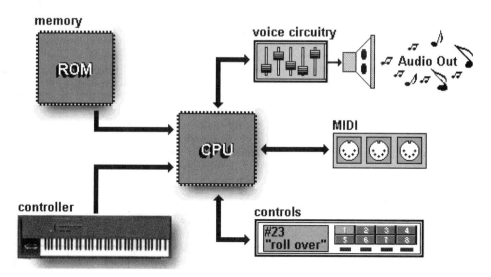

Figure 4.1   The basic components of an electronic musical instrument.

- **Performance controllers:** These include such controllers as keyboards, drum pads, and wind controllers for inputting performance data directly into the electronic instrument in real time or for transforming a performance into MIDI messages. Not all instruments have a built-in controller. These devices (commonly known as modules) contain all the necessary processing and sound generating circuitry; however, they can save lots of space in a cramped studio by eliminating redundant keyboards or other controller surfaces.

- **Control panel:** The all-important human interface of data entry controls and display panels lets you select and edit sounds, route and mix output signals, and control the instrument's basic operating functions.

- **Memory:** Memory is used for storing important internal data (such as patch information, setup configurations, and/or digital waveform data). This digital data can be in the form of either read-only memory (ROM; data that can only be retrieved from a factory-encoded chip, cartridge, or CD-ROM) or random access memory (RAM, memory that can be stored onto or retrieved from a memory chip, cartridge, hard disk, or recordable optical media).

- **Voice circuitry:** Depending on the device type, this section can either generate analog sounds (voices) or it can be used to instruct digital samples that are permanently recorded into memory to be played back according to a specific set of parameters. In short, it's

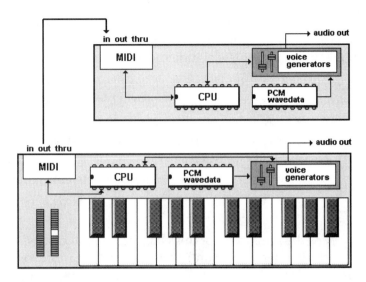

Figure 4.2 Communication among instruments via MIDI.

used to generate or reproduce a sound patch, which can then be amplified and heard via speakers or headphones.

- **Auxiliary controllers:** These are external controlling devices that can be used in conjunction with an instrument or controller. Examples of these include foot pedals (providing continuous-controller data), breath controllers, and pitch-bend or modulation wheels. In addition, certain controllers will only let you switch a function on and off. Examples of such controllers are sustain pedals and vibrato switches.

- **MIDI communications ports:** These are used to transmit and/or receive MIDI data.

Generally, no direct link is made between each of these functional blocks; the data from each of these components is routed and processed through the instrument's CPU. For example, should you wish to select a certain sound patch from the instrument's control panel, the CPU would recall from memory all of the waveform and sound-patch parameters that are associated with that particular sound. These parameters will then modify the internal voice circuitry, so that when a key on the keyboard is pressed, the sound generators will output the patch at a frequency that's equivalent to the appropriate note value.

Most, if not all, of an instrument's components can be accessed or communicated amongst several devices (in one form or another) via MIDI. In fact, MIDI can be used in many ways to communicate performance, patch and system setup information, and even audio data between devices. A few of the many examples include the following:

- Playing a synthesizer's keyboard could transmit performance messages to a synth module (Figure 4.2).
- All of an instrument's sound-patch data settings could be transmitted to or received from a sequencer or other instrument as MIDI system exclusive messages (SysEx).
- Sampled audio could be exchanged between samplers using the MIDI sample dump standard or SCSI (a computer protocol that communicates data at high speeds).

For the remainder of this chapter, we will discuss the various types of MIDI instruments and controller devices that are currently available on the market. These instruments can be grouped into such categories as keyboards, percussion, MIDI guitars and strings, woodwind instruments, and controlling devices.

Keyboards

By far, the most common instruments that you'll encounter in almost any MIDI production facility belong to the keyboard family. This is due, in part, to the fact that keyboards were the first electronic music devices to gain wide acceptance, and that MIDI was initially developed to record and control many of their performance and control parameters. The two basic keyboard-based instruments are the synthesizer and the digital sampler.

The Synthesizer

A *synthesizer* (or *synth*) is an electronic instrument that uses multiple sound generators to create complex waveforms that can be combined (using various waveform synthesis techniques) into countless sonic variations. These synthesized sounds have become a basic staple of modern music and range from those that sound "cheesy," to those that closely mimic traditional instruments, all the way to those that generate other-world, ethereal sounds that literally defy classification.

Synthesizers (Figure 4.3) generate sounds using a number of different technologies or program algorithms. The earliest synthesizers were analog in nature and generated sounds using a technology known as frequency modulation (FM) synthesis. Today, however, modern FM synthesis is usually implemented entirely in the digital domain.

Figure 4.3
The Minimoog. (Courtesy of Moog Music Inc., www.moogmusic.com)

FM synthesis techniques generally make use of at least two signal generators (commonly referred to as "operators") to create and modify a voice. Often, this is done by generating a signal that modulates or changes the tonal and amplitude characteristics of a base carrier signal. More sophisticated FM synths use up to four or six operators per voice, and these generators will also often use filters and variable amplifier types to alter a signal's characteristics into a sonic voice that either loosely imitates acoustic instruments or creates sounds that are totally unique.

Another technique used to create sounds is *wavetable synthesis*. This technique works by storing small segments of digitally sampled sound into a read-only memory chip. Various sample-based synthesis techniques use sample looping, mathematical interpolation, pitch shifting, and digital filtering to create extended and richly textured sounds that use a very small amount of sample memory.

These sample-based systems (Figure 4.4) are often called *wavetable synthesizers* because a large number of prerecorded samples are encoded within the instrument's memory and can be thought of as a "table" of sound waveforms that can be looked up and used when needed. Once selected, a

Figure 4.4
The Korg Trinity workstation. (Courtesy of Korg USA, Inc., www.korg.com)

range of parameters (such as wavetable mixing, envelope, pitch, volume, pan, and modulation) can be modified to control a sample's overall sound character.

Synthesizers come in all shapes and sizes and use a range of patented synthesis techniques for generating and shaping complex waveforms into every type of sound imaginable. They can be used to provide a vast palette of sounds and textures that can be played over the keyboard's range in a polyphonic fashion using 16, 32, or even 64 simultaneous voices. In addition, these synths often include a percussion section that can play a full range of drum sounds, as well as acoustic, electronic, and European-style percussion setups. Reverb and other effect types are also built into the architecture of most modern-day synths, which reduces the need for using extensive outboard effects.

The Synthesizer Module

Synthesizers are also commonly designed into 19-inch rack-mountable or table-top systems (Figures 4.5–4.7). These devices, which are also known as *synth modules* or *expanders*, often contain all of the features of a standard synthesizer, except they don't have a keyboard controller. This space-saving feature means that more synths can be placed into your system and can be

Figure 4.5
Basic MIDI system that incorporates a synth module.

Figure 4.6
"Orbit V2, the Dance Planet" sound module. (Courtesy of E-mu Systems, Inc., www.emu.com)

Figure 4.7
Roland JV-2080 synth module. (Courtesy of Roland Corp. US, www.rolandus.com)

controlled from a master keyboard controller or sequencer without cluttering your system up with redundant keyboards.

Soundcard Synths

By far, the greatest number of installed synthesizers are those that have been designed into generic PC soundcards. These devices (which can be found in almost every home) are often designed into a single chip set, and often generate sounds using a simple form of digitally controlled FM synthesis. Although more expensive soundcards will often use wavetable synthesis to create richer and more realistic sounds, both card types will almost always conform to the General MIDI specification, which has universally defined the overall patch and drum-sound structure so that MIDI files will be uniformly played by all synths with the correct voicings and levels. Further information about General MIDI can be found in Chapter 9.

Note that some soundcard types aren't shipped with a synthesizer; instead they are equipped with a standard plug-in expansion port. These hardware cards (know as daughterboards) contain a synth chip set that can be easily inserted into the soundcard's circuitry.

Software Synthesis and Sample Resynthesis

Because wavetable synthesizers derive their sounds from prerecorded samples that are stored in a digital memory media (often ROM chips), it logically follows that these sounds can also be stored on hard disk (or any other media), which can then be temporarily loaded into the RAM memory of a personal computer. This process of downloading wavetable samples into a computer and then manipulating these samples is used to create what is known as a virtual or software synthesizer (Figures 4.8 and 4.9). These systems are often capable of generating sounds or importing existing digital audio samples and then using various modular algorithms to change these sounds into almost any texture or synthesized sound that you could possibly imagine.

In short, modular software synthesis works by linking various signal processing modules in a chain or parallel fashion to generate or modify a

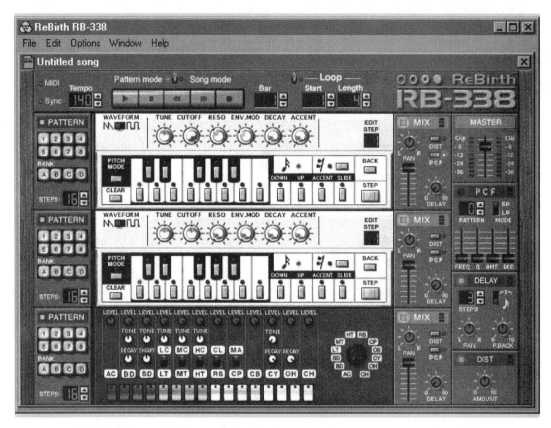

Figure 4.8 ReBirth software synthesizer. (Courtesy of Steinberg N. America, www.steinberg-na.com)

sound. These modules consist of such traditional synthesis building blocks as oscillators, voltage-controlled amplifiers, voltage-controlled filters, and mixers to modify, combine, and attenuate the signal's overall harmonic content structure. Because the system exists in software, a newly created sound patch can be saved to disk for later recall.

As you might expect, the depth and capabilities of a software synth depend on the quality of the program and its generation techniques, wavetable signal quality, sample rate, and overall processing techniques. These can range from a simple General MIDI software synthesizer (like those that are currently being designed for Microsoft Windows 95, 98, and NT operating systems) to professional level software synthesizers that let you import, edit, and combine wavetable sound data with an amazing degree of ease and control.

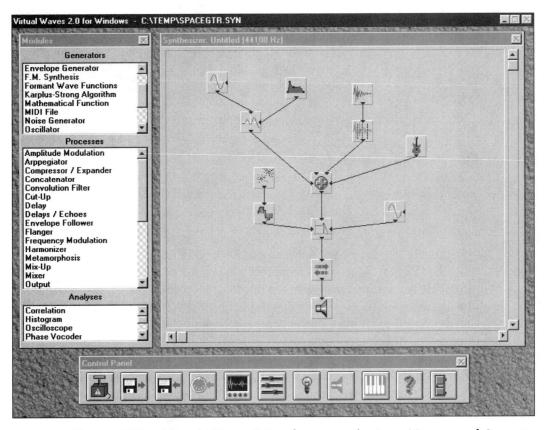

Figure 4.9 Virtual Waves 2.0 software synthesizer. (Courtesy of Synoptic, www.synoptic.net)

System Exclusive: "The Musician's Pal"

It's simply impossible for me to talk about synthesizers and not bring up one of an electronic musician's greatest tools: system exclusive. As we saw near the end of Chapter 2, the system exclusive message (or *SysEx* for short) makes it possible for MIDI manufacturers, programmers, and designers to communicate customized MIDI messages among MIDI devices (Figure 4.10). The general idea behind SysEx is that it uses MIDI messages to transmit and receive program/patch data or real-time parameter information from one device to another. It's sort of like having a synth that's a digital chameleon. One moment the synthesizer is configured with a certain set of sound patches and setup data and then, after having received a new SysEx data dump, you could easily end up with a synth that's literally full of new

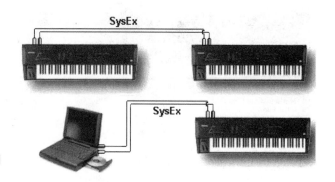

Figure 4.10
SysEx patch data can be transmitted between identical devices or between a SysEx program and a MIDI device.

and exciting (or not-so-exciting) sounds and settings. In the following paragraphs we look at a few examples of how SysEx can be put to good use.

SysEx can be used to transmit patch and overall setup data between identical makes and (most often) models of synthesizers. Let's say that you have a Brand X Model Z synthesizer and it turns out that you have a buddy across town who also has a Brand X Model Z. That's cool, except your buddy's synth has a completely different set of sound patches—and you want them! SysEx to the rescue! All you need to do is go over and transfer your buddy's patch data into your synth (to make life easier, make sure you take your instruction manual along, just in case you run into a snag) and follow these simple guidelines:

1. Back up your present patch data! This can be done by transmitting a SysEx "dump" of your synthesizer's entire patch and setup data to disk, SysEx utility software (often shareware), or to your MIDI sequencer. This is so important that I'll say it again: Back up your present patch data before attempting a SysEx dump! If you forget and download a SysEx dump, your previous settings will be lost until you contact the manufacturer or take your synth back to your favorite music store to reload the data.
2. Do the data transfer according to the manufacturer's manual.
3. Back up your new patch data as per the previous backup!

Now that you have both your own and your friend's patch data backed up as a SysEx data dump, you can reload your synth (using any of the above transfer methods). This will effectively give you access to lots more sounds, which will be limited solely by the number of SysEx dumps that you have hoarded for that particular synth.

SysEx also lets you get patch data from the web. One of the biggest repositories of SysEx data is the Internet. To surf the web for SysEx patch data, all you need to do is log on to your favorite search engine site (my current personal favorite is www.yahoo.com), enter the name of your synth, and hit return. You'll be surprised at how many "hits" you'll come across, many of which are chock-full of SysEx dumps that can be downloaded for transmission into your synth.

I should point out that SysEx data that's grabbed from the web, disk, or any other format will often be encoded using several file format styles (unfortunately, none of which are standardized). A SysEx dump that was encoded using sequencer Z might not be recognized by sequencer Y. For this reason, many dumps are encoded using SysEx utilities that have been semi-standardized by virtue of their sheer numbers of users.

Before we move on to another subject, I'd like to stop for a moment and give you some advice that just might save your butt when an instrument's battery decides to give out, your main hard drive goes belly-up, or the instrument's memory locks up (it's not common, but it's common enough that it happened to me):

1. *ALWAYS* dump the entire SysEx global settings (patch data, setup, etc.) to your patch editor and/or your sequencer's SysEx dump utility soon after getting a new device or instrument. This gives you a copy of all the settings that were programmed into it at the factory (or by the previous user, in the case of used instruments).
2. Store a copy of this dump with your SysEx backup files. (*ALWAYS* make backups, preferably a couple.)
3. *ALWAYS* save newly edited patch banks or banks that come to you from any source (i.e., friends, commercial disks, the Internet) into your patch editor and/or your sequencer's SysEx dump utility and then back them up.

What's the moral of this story? Always have a full SysEx dump of all your instruments on file (this is probably best done from your sequencer's SysEx utility) and keep fresh backups of these files. If you're a working music stiff, I can almost promise that the time will come when you'll need them.

The Digital Sampler

A *digital sampler* (Figures 4.11 and 4.12) is a device that's capable of converting a segment of audio into a digitized form and then loading this digital data into its internal RAM memory for playback in a polyphonic, musical fashion. Basically, a sampler can be thought of as a wavetable synthesizer that lets you record, edit, and reload the samples into its internal memory. Once loaded, these sounds (whose length and complexity is limited only

Figure 4.11
The BOSS Dr.Sample SP-202 sample. (Courtesy of Roland Corp. US, www.rolandus.com)

Figure 4.12
The Ensoniq ASR-X sampler. (Courtesy of Ensoniq Corp., www.ensoniq.com)

by RAM size and your imagination) can be looped, modulated, filtered, and amplified according to user or factory setup parameters so as to modify a sample's overall waveshape and envelope. Signal processing capabilities, such as basic editing, looping, gain changing, reverse, sample-rate conversion, pitch change, and digital mixing capabilities are also often designed into the system.

A sampler's design will often include a keyboard or set of trigger pads that let you polyphonically play samples as musical chords, sustain pads, triggered percussion sounds, or sound effect events. These samples can be played according to the standard Western musical scale (or any other scale, for that matter) by altering the playback sample rate. Pressing a low-pitched key on the keyboard will cause the sample to be played back at a low sample rate, while pressing a high-pitched one will cause the sample to be played back at a rate that would put Mickey Mouse to shame. By choosing

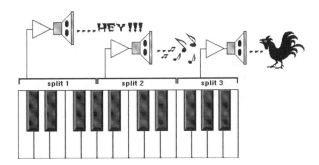

Figure 4.13
Samples can be mapped to various "zones" on a keyboard.

the proper sample-rate ratios, samples can be simultaneously played at various pitches that correspond to standard music intervals.

Once a set of samples has been recorded or recalled from disk, each sample in a multiple-voice system can be split (using a process known as *mapping*) across a performance keyboard (Figure 4.13). In this way, a sound can be "mapped" to a specific "zone" or range of notes and/or into velocity "layers," allowing the system to be programmed so that various key pressures will trigger two or more different samples. For example, a single key on a loaded piano setup might be layered to trigger two samples. Pressing the key at velocities ranging from 0 to 64 might sound a sample of a piano note that was played softly, while a velocity range of 65 to 127 might output a sample of a piano note that was struck hard. Mapping can also be used to create wild soundscapes that change not only with the played keys, but with different velocity as well.

A professional sampler often includes such features as integrated signal processors (which can be assigned to all or individual samples), multiple outputs (offering isolated channel outputs for added mixing and signal processing power, or for recording individual voices to a multitrack tape recorder), and integrated MIDI sequencing capabilities.

Distributing data to and from a sampler can be done in many ways. Almost every sampler has a floppy or fixed/removable hard drive for saving waveform or system data and many are equipped with a port for adding a CD-ROM drive. Distribution between samplers or to/from a sample editing program can be handled through the use of the slower MIDI sample-dump standard, or via a high-speed SCSI (small computer system interface) port. Detailed information on these data protocols can be found in Chapter 8.

The Sampler Module

As with synthesizers, some samplers (known as *sample modules*; Figures 4.14 and 4.15) integrate all of their necessary signal processing, programming, and digital control structures into a single, 19-inch rack-mountable

Figure 4.14
The Akai S2000 sampler. (Courtesy of Akai Musical Instrument Corp., www.akai.com.)

Figure 4.15
The ESI-4000 digital sampler. (Courtesy of E-mu Systems, Inc., www.emu.com.)

unit. Because these devices don't have a keyboard, they must be controlled from an external controller (such as an external keyboard controller, drum pads, sequencer or other controller type).

Sample Disks, CD-ROMs, and Other Media: "The Musician's Other Pal"

Just as patch data in the form of SysEx dump files have the effect of breathing new life into your synthesizer, sample data that has been preformatted for your sampler is another consumable favorite. These sample files (which

Figure 4.16
Sample collections on CD-ROM. (Courtesy of East/West Communications, Inc., www.soundsonline.com)

can be obtained from a number of sources) are commonly preformatted by a professional musician/programmer to contain all the necessary loops, system commands, and sound generator parameters so that all you "ideally" need to do is load the sample and begin having fun.

It almost goes without saying that individual samples and small sample libraries can be downloaded from the Internet. However, larger and more complex libraries are far too data intensive to download and are often commercially created by companies for distribution on CD-ROM (Figure 4.16).

The MIDI Keyboard Controller

The *MIDI keyboard controller* is a keyboard device that's expressly designed to control other devices within a connected MIDI system. It contains no internal tone generators or sound-producing elements. Instead, its design often includes a high-quality weighted performance keyboard and a wide range of controls for handling MIDI performance, control, and device switching events.

Often a controller's keyboard will offer an extended key range (66 or 88 keys) that's weighted to mimic the action of a grand piano keyboard. Full mapping implementation allows the keys to be split into user-defined zones that can be programmed to overlap with each other, to occupy the same area, or to cover separate sections of the keyboard. Each zone can be assigned to its own MIDI channel, program-change number, continuous-controller functions (which can be assigned to a modulation wheel, pitch-bend wheel, or data fader), as well as have its own aftertouch, velocity curves, and key transposition settings. These MIDI channel mapping assignments make it possible for multiple instruments, voicings, and MIDI parameters to be mapped to various zones over the playing surface. Once programmed, the overall parameter settings can be stored to create an overall system snapshot of MIDI channels, program changes, key ranges, etc.

The daddy of the keyboard controller set is the MIDI grand piano. Some of these are a fully functional grands, while others have the look, but don't include strings. Kits are also available that let you retrofit an existing piano with a MIDI controller kit.

Depending on their design, MIDI grands may use traditional electrical contact or fiber optic technology to capture keyboard expression accurately, without interfering with the normal touch response of the piano's action. In addition to having pitch-bend and modulation wheels, the sustain and una corda pedals on most MIDI grands can be used as controllers to sustain or soften the sound of external MIDI instruments.

Non-Keyboard Controllers

On certain occasions, more nontraditional controller types can be used in place of a keyboard or to augment a keyboard's functionality. For example, most drum machines are able to transmit MIDI note on/off and velocity messages. In a pinch, these can be programmed to control any number of MIDI device types. Alternately, control commands, program changes, and other switching functions can also be done on the fly by assigning program-change values to devices like a MIDI-fied guitar effects foot pedal. In short, never forget that necessity is often the mother of invention. Problems relating to MIDI control can often be dealt with in interesting and cost-effective ways.

Percussion

One of the first applications in sample technology was to record drum and percussion sounds, making it possible for electronic musicians (mostly keyboard players) to add percussion samples to their own compositions. Out of this sprang a major class of sample and synthesis technology that lets an artist create drum and percussion sounds by using their synths, drum machines, or samplers.

Over the years, MIDI has brought sampled percussion within the grasp of every electronic musician, from those whose performance skills range from that of the frustrated drummer to professional percussionist/ programmers who use their skills to perform live or to build up complex sequenced drum patterns.

The Drum Machine

The *drum machine* (Figures 4.17 and 4.18) is most commonly a sample-based digital audio device that can't record audio into its internal memory. Instead it uses ROM-based, prerecorded waveform samples to reproduce high-quality drum sounds. These factory-loaded sounds often include a wide assortment of drum sets, percussion sets, rare and wacky percussion hits, and effected drum sets (i.e., reverberated, gated, etc.). Who knows, you might even encounter scream hits by the venerable King of Soul, James Brown. These prerecorded samples can be assigned to a series of playable keypads that are generally located on the machine's top face, providing a straightforward controller surface that often sports velocity and after-touch dynamics. Drum voices can be also assigned to each pad and edited using control parameters such as tuning, level, output assignment, and panning position.

Figure 4.17
The Akai MPC-2000 MIDI production center. (Courtesy of Akai Musical Instrument Corp., www.akai.com.)

Figure 4.18
The Roland MC-303 groove box. (Courtesy of Roland Corp. US, www.rolandus.com)

The assigned sounds can be played live or from a programmed sequence track. Alternately, certain drum machines have a built-in sequencer that has been specifically designed to arrange drum/percussion sounds into a rhythmic sequence (known as a *drum pattern*). These patterns often consist of basic variations on a rhythmic groove, or they can be built from patterns taken from an existing library of playing styles (such as rock, country, or jazz). Drum machines that have this feature will often let you chain these patterns together into a continuous song. Once a song is assembled, it can be played back using an internal MIDI clock source, or it can be synchronously driven from another device (such as a sequencer) using an external MIDI clock source.

Although a number of drum machine designs include a built-in sequencer, it's more likely that these studio workhorses will be triggered from a MIDI sequencer. This lets you take full advantage of the real-time performance and editing capabilities that a sequencer has to offer. For example, sequenced patterns can be easily created in step time (where notes are entered and assembled into a rhythmic pattern one note at a time) and then linked into a song that's composed of several rhythmic variations. Alternately, drum tracks can be played into a sequencer on the fly, creating a live feel, or you can merge step- and real-time tracks together to create a more human-sounding composite rhythm track. In the final analysis, the style and approach to composition is entirely up to you.

Most drum machine designs include multiple outputs that let you route individual or groups of voices to a specific output on a mixer or console. This feature allows these isolated voices to be individually processed (using equalization, effects, etc.) or to be recorded onto separate tracks of a multi-track tape recorder.

Alternative Percussion Voices

In addition to the numerous sounds that can found in a drum machine, a virtually unlimited number of percussion sounds can be obtained from other sources. As was mentioned earlier, a synthesizer will often include several drum and/or percussion setups that are often mapped over the entire keyboard surface. Sampler libraries will almost always include a never-ending number of percussion instruments and drum sets. Sound files can be loaded into a hard disk editor to build up rhythm tracks, or you can "lift" percussion loops from loop disks that are available on CD. Again, the sky's only limited by your imagination.

I'd also like to take this time to introduce you to the concept of recording your own samples. If you own a sampler and can't find the sounds that you want, then sample your own! It's a surefire way to personalize your music and/or effects in a way that will often cause everyone's ears to perk up.

MIDI Drum Controllers

MIDI drum controllers are used to translate the voicings and expressiveness of a percussion performance into MIDI data. These devices are great for capturing the feel of a live performance, while giving you the flexibility of automating or sequencing a live event.

It has long been a popular misconception that MIDI drum controllers have to be expensive. Personally, I found a simple controller in a toy store that has four pads, velocity scaling, and basic MIDI implementation for $69! Beyond that, there are still tons of ways to play and create patterns in your studio or on stage. The following sections outline some of the more common examples.

Drum Machine Keypads

One of the more straightforward of all drum controllers is the drum pads that are designed into most drum machines. By calling up the desired setup and voice parameters, you can then go about the business of playing your performance directly to a sequenced track. It's also a simple matter to trigger other devices from these keypads. For example, you could assign the pads to a channel that's being responded to by your favorite synth or sampler, and then trigger these sounds directly.

Coming under the "Don't try this at home" category, these controller pads are generally too small and not durable enough to withstand drumsticks or mallets. For this reason, they're generally played with the fingers. In addition, their velocity sensitivity is often limited to a range of sensitivity steps and doesn't express the full range of velocity levels.

The Keyboard as a Percussion Controller

Because drum machines respond to external MIDI data, controllers such as a MIDI keyboard can be used to trigger drum machine voices. One advantage of playing percussion sounds from a keyboard is that sounds can be triggered more quickly because the playing surface is designed for fast finger movements and doesn't require full hand/wrist motions. Another advantage is the ability to express velocity over the entire range of possible values (0–127), instead of the limited velocity steps that are often available on drum pads.

Most drum machines let you manually assign drum and percussion voices to a particular MIDI note value. Because the percussion sounds may not be related to any musical interval, this means that you can assign the drum voices to any keyboard note and range that you'd like. In addition, these drum sounds can be assigned to a particular range of notes over a split keyboard arrangement, allowing other sound patches to be simultaneously addressed on the same playing surface.

Drum Pad Controllers

In more advanced MIDI project studios or live stage rigs, it's often necessary for a percussionist to have access to a playing surface that can be played like a real percussion instrument. At these times, a dedicated *drum pad controller* (Figure 4.19) would be perfect for the job. Drum controllers often have between six and eight playing pads, which are generally large and can be mapped to various percussion samples with regard to velocity, the area that is struck, etc. Their playing surface can be played with either the fingers, hands, percussion mallets, or drumsticks.

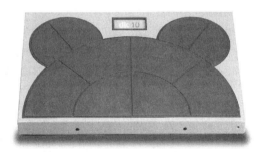

Figure 4.19
The drumKAT MIDI dk-10
drum controller. (Courtesy
of E-mu Systems, Inc.,
www.emu.com.)

Drum pad controllers can also be programmed to give you fast and easy access to any number of setup parameters. This lets you assign each pad to a full range of MIDI parameters (such as MIDI channels, velocity levels, one or more note numbers, after touch, etc.), save them, and then switch between the program setups during a sequence or live performance.

MIDI Drums

We could take the idea of a realistic playing surface a step further by including those controllers that are physically configured to resemble an actual drum trap set (Figure 4.20). These setups let you have all the functionality and the animal magnetism of having a real drum set, while having all the advantages (like recordability, automation, controller, and effects switching) that MIDI has to offer. Besides, you could play your traps over headphones at 2 A.M. without disturbing the neighbors.

Taking realism even further, you'll be happy to know that real drums can be MIDI-fied without too much fuss. This is done by miking the drum (or by using a contact pickup) and using that signal as a trigger for outputting the proper MIDI messages to a drum machine or sampler (Figure 4.21). Certain drum machines and other dedicated trigger devices have line and low-level trigger inputs that can be used to transmit MIDI messages from a live event.

Figure 4.20
The Roland V-Drums. (Courtesy of Roland Corp. US,
www.rolandus.com)

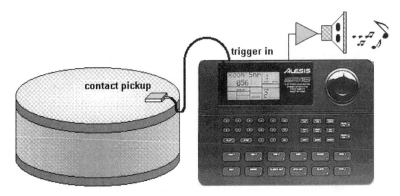

Figure 4.21   A pickup signal can be used to trigger a MIDI event from a live source.

Figure 4.22
The MalletKAT MIDI vibra-
phone. (Courtesy of E-mu
Systems, Inc.,
www.emu.com.)

Because the trigger source is an electrical audio signal, the original source can be almost anything. For example, you could mike a snare drum in the studio and use that signal to trigger a monster snare. Alternately, you could use the already-recorded snare track as a trigger source. This type of production work definitely comes under the "use your imagination" category.

The MIDI Vibraphone

Unlike drum machine pads or keyboard controllers, the *MIDI vibraphone* (Figure 4.22) is generally used by professional percussionists. These vibes are commonly designed with a playing surface that can be fully configured using its internal setup memory to provide for user-defined program changes, playing-surface splits, velocity, after touch, modulation, etc.

MIDI Guitars

Guitar players often work at stretching the vocabulary of their instruments beyond the norm. They love doing nontraditional gymnastics using such tools of the trade as distortion, phasing, echo, feedback, etc. Due to advances

Figure 4.23
The MA-3 Vintage Elite
MIDI guitar. (Courtesy of
Virtual DSP Corp.,
www.midiaxe.com)

in guitar pickup and microprocessor technology, it's now possible for the notes and minute inflections of guitar strings to be accurately translated into MIDI data (Figure 4.23). With this innovation, many of the capabilities that MIDI has to offer are now available to the electric (and electronic) guitarist. For example, a guitar's natural sound can be layered with a synth pad that has been transposed down to give a rich, thick sound that just might shake your boots. Alternately, recording a sequenced guitar track in sync with the audio tracks would give a producer the option of changing and shaping the sound in mixdown. On-stage program changes are also a big plus for the MIDI guitar. These let the player radically switch between guitar voices from the synth or sequencer or by simply stomping on a MIDI foot controller.

MIDI Wind Controllers

MIDI wind controllers (Figure 4.24) differ from keyboard and drum controllers because they are expressly designed to bring the breath and key articulation of a woodwind or brass instrument to a MIDI performance. These controller types are used because many of the dynamic- and pitch-related expressions (such as breath and controlled pitch glide) simply can't be communicated from a standard music keyboard. In these situations, wind controllers can often help create a dynamic feel that's more in keeping with their acoustic counterparts by using an interface that provides special touch-sensitive keys, glide- and pitch-slider controls and sensors for outputting real-time breath control over dynamics.

Figure 4.24
The EWI 3020 electronic
wind instrument. (Courtesy
of Akai Musical Instrument
Corp., www.akai.com)

5

SEQUENCING

Apart from electronic musical instruments, one of the most important tools that can be found in the modern-day project studio is the *MIDI sequencer*. Basically, a sequencer is a digital device that's used to record, edit, and output MIDI messages in a sequential fashion. These sequential messages are generally arranged in track-based format that follows the modern production concept of having separate instruments (and/or instrument voices) located on separate tracks. This traditional interface makes it easy for us humans to view MIDI data as though they were tracks on a tape recorder that follows along a straightforward linear time line (Figure 5.1).

These "tracks" contain MIDI-related performance and control events that are made up of channel and system messages such as note-on, note-off, velocity, modulation, after touch, and program/continuous-controller messages. Once a performance has been recorded into a sequencer's memory, these events can be graphically or audibly edited into a musical performance. The data can then be saved to any digital storage media and recalled at any time, allowing the data to be played back in its originally recorded order.

As mentioned before, most sequencers are designed to function in a way that's similar to their distant cousin, the multitrack tape recorder. This gives us a familiar operating environment, whereby each instrument, set of layered instruments, or controller data can be recorded onto separate, synchronized tracks. Like its multitrack cousin, each track can be rerecorded, erased, copied, and varied in level during playback. However, because the recorded data is digital in nature, a MIDI sequencer is far more flexible in its editing speed and control in that it offers all the cut-and-paste, signal

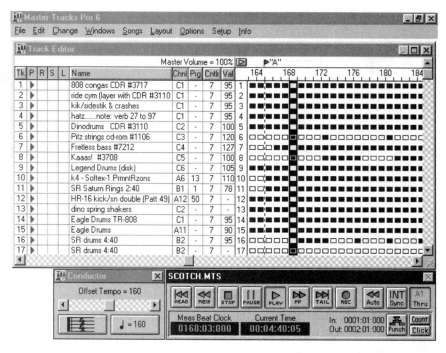

Figure 5.1 Basic track and time line layout of a MIDI sequencer. (Courtesy of Passport Designs, Inc., www.passportdesigns.com)

processing, and channel routing features that we've come to expect from a digital production device.

The number of tracks that a sequencer offers vary from one manufacturer and model type to the next and can range from eight to practically infinity. While basic hardware and software sequencers can record and assign data on up to 16 simultaneous MIDI channels, most professional sequencing software packages can easily assign a MIDI track to any channel on two, eight, or more individual input/output ports. This means that most newer software packages can easily transmit data over 32, 128, or more independent MIDI channels.

Hardware Sequencers

Hardware sequencers are stand-alone devices that are designed for the sole purpose of sequencing MIDI data. These systems include a specially designed CPU and operating system, memory, MIDI ports, and integrated controls for performing sequence-specific functions.

Ease of use and portability are often the advantages of a hardware sequencer, most of which are designed to emulate the basic functions of a tape transport (record, play, start/stop, fast forward, and rewind). Generally, a moderate number of editing features are available, including note editing, velocity and other controller messages, program change, cut-and-paste and track merging capabilities, and tempo changes.

Hardware sequencers commonly use an LCD to display programming, tracking, and editing information. This display type is often small in size and resolution and is generally limited to information that relates to one parameter or track at a time.

Integrated Workstation Sequencers

Some of the newer and more expensive keyboard synthesizers and samplers are designed with a built-in sequencer. These portable systems (which are commonly called *workstations*) have the advantage of letting you take your instrument and sequencer on the road without having to drag along your whole system.

Integrated sequencers often have the disadvantage of not offering an extensive range of editing tools beyond transport functions, punch-in/out commands and other basic edit functions. However, for basic sequence work, they are often more than adequate and can be used with other instruments that are connected in a MIDI chain.

Software Sequencers

By far, the most common sequencer type is the software sequencing program. These programs run on all types of personal computers and take advantage of the hardware and software versatility that a computer can offer in the way of speed, hardware flexibility, digital signal processing, memory management, and signal routing.

A computer-based sequencer offers several advantages over its hardware counterpart. Among these are increased graphics capabilities (which let you have direct control over track and transport-related functions), standard computer cut-and-paste techniques, a windowed graphic environment (allowing easy manipulation of program and edit-related data), routing MIDI to multiple ports in a system, and graphic assignment of instrument voices via program change messages, not to mention the ability to save and recall files using standard computer memory media.

Basic Introduction to Sequencing

When dealing with any type of sequencer, one of the most important concepts to grasp is that these devices don't store sound directly; instead, they encode MIDI messages that instruct instruments as to what note is to be played, over what channel, at what velocity, and at what, if any, optional controller values. In other words, a sequencer simply stores command instructions that follow in a sequential order. These instructions tell instruments and/or devices how their voices are to be played or controlled. This means that the amount of encoded data is a great deal less memory intensive than its hard disk audio or video recording counterparts. Because of this, the data overhead that is required by MIDI is very small. A computer-based sequencer can work simultaneously with the playback of digital audio tracks, video images, Internet browsing, etc., all without unduly slowing down a computer's CPU. For this reason, MIDI and the MIDI Sequencer is an ideal media form for the computer.

As you might expect, a lot of hardware, integrated, and software sequencers are currently on the market. Each type and model offers a unique set of advantages and disadvantages. It's also true that each sequencer has its own basic operating "feel" and, as a result, choosing one over another is totally up to personal preference.

As with anything, when buying a sequencer, you should shop carefully, keeping your personal working habits and future growth needs in mind and—most of all—choose the system or version that's right for you. On the hardware side, this means spending some time at your favorite music store, while on the software side, you might want to visit various manufacturers' web sites and take the time to check out their latest demos.

Now that we've had a basic overview, let's take a closer look at some of the basic functions this important tool of the trade has to offer. Before we begin, I should point out that exhaustively covering all of the features of every sequencer would take a few hundred pages. Therefore, by necessity, this section merely presents an overview. For those who already have the sequencer of your dreams (or worst nightmares), I strongly recommend that you read and reread its manual in order to reap the full benefits of the system and then experiment to get the most out of what it has to offer.

Recording

Commonly, a sequencer is used as a digital workspace for creating personal compositions in environments that range from a bedroom to more elaborate project studios. Whether hardware or software based, most sequencers use a working interface that's designed to emulate the traditional multitrack recording environment. A tape-like transport lets you move from one location to the next using standard play, stop, fast forward, rewind, and record command buttons. Beyond using the traditional record-enable button(s) to select the track(s) that you want to record onto, all you need to do is select the MIDI output port (if more than one exists), MIDI channel, instrument patch, and other setup information. Then press the record button and begin playing.

Once you've finished laying down a track, you can jump back to the beginning of a sequence and listen to your original track while continuing to lay down additional tracks until a song begins to form.

Although only one MIDI track is usually recorded at a time, most mid- and professional-level sequencers let you record more than one track at a time. This multitrack feature makes it possible for several performers to record to a sequence in one live pass. It can also let you transfer all the channels of a completed sequence from one sequencer to another in real time.

Almost all sequencers are capable of *punching in* and *out* of record while playing a sequence (Figure 5.2). This commonly used function lets you drop in and out of record on a selected track (or series of tracks) in real time. Although punch-in/out points can often be manually performed on the fly, most sequencers can automatically perform a "punch." This is generally done by graphically or numerically entering in the measure numbers that mark the punch-in/out location points. Once done, the sequence can be rolled back to a point a few measures before the punch-in point and the artist can play along while the sequencer automatically performs the switching functions.

In addition to recording a performance one track at a time or all at once, most sequencers will let you enter note values into sequence one note

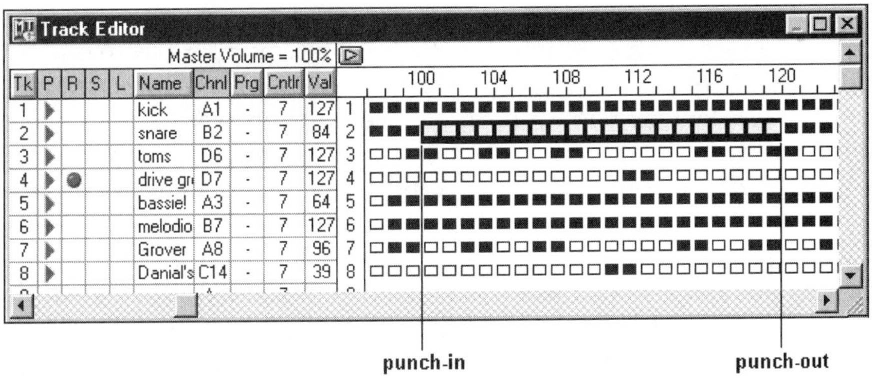

Figure 5.2 Sequencer track window showing punch-in and punch-out points.

at a time. This feature (known as *step time*), lets you enter and edit a rapid or difficult passage into a sequence with little or no difficulty.

The way that step time is entered into a sequence varies from sequencer to sequencer. By giving the sequencer a basic tempo and note length (e.g., 1/4 note, 1/16 note) and then manually entering the notes from a keyboard or other controller, you can continue to enter notes of a specified duration into the sequence, until a rest value or new note of a different duration value is selected. This data entry style is often (but not always) used with fast, high-tech and dance styles, where real-time entry just isn't possible or accurate enough for the song.

Whether you're recording a track in real time or in step time, you'll often want to select the proper song tempo before recording a sequence. I mention this because most sequencers are able to output a *click track* that can be used as an accurate, audible guide for keeping time with the song's selected tempo.

A click track can be set to sound a beep or a MIDI note on the measure boundary, or (for a more accurate timing guide) on the first beat boundary and on the measure divisions (i.e., tock, tick, tick, tick, tock, tick, tick, tick, etc.). Most sequencers can output a click track by either using a dedicated beep sound (often from the PC's internal speaker) or by sending note-on messages to one or more MIDI instruments in the system. The latter lets you use any sound you want and often at definable velocity levels. For example, a kick could sound on the beat, while a snare sounds out the measure divisions. A strong reason for using a click track (at least initially) is that its rhythmic accuracy will let you correct for timing problems that might otherwise occur during the performance. For freeform compositions, you might decide not to use a click at all.

Editing

One of the more important features a sequencer has to offer is its ability to edit sequenced tracks. Editing functions and capabilities often vary from one sequencer to the next, with certain hardware systems offering only the most basic cut-and-paste, signal processing, and data routing capabilities. Most software sequencers, on the other hand, generally offer a wide range of editing functions that can manipulate MIDI data in a simple, easy-to-use graphic environment.

As was mentioned, hardware sequencers will often display entry- and editing-related information using an LCD screen that's often 2 inches by 5 inches or smaller. As a result, the way in which parameters are edited can be somewhat limited by the display type. To some, this simplifies the editing process by requiring that you do much of your editing by ear, with entry and edit resolutions being entered at the measure level, instead of at the note and clock resolution level.

Software sequencers, on the other hand, display all of their transport, control, and editing functions on the much larger surface of your computer monitor. Because software can be easily reconfigured to best deal with the task at hand, these sequencers can graphically tailor their display and control surface in a manor that's intuitive and easy to use (at least that is the theory).

The remainder of this section outlines the various control and editing features that can be found in current sequencing packages. For practicality's sake, we'll be looking at examples and general functionality from a software point of view, using screenshots from various manufacturers as a visual guide. Although hardware sequencers often vary in their overall capabilities, many of the following examples can be applied to these devices as well.

Main Edit Screen

The main screen of a sequencer (Figure 5.3) will generally include a transport control surface, track editing and navigation screen, and a time signature/tempo window. As already mentioned, the transport gives us access to the basic play, stop, fast forward, rewind, and record buttons, while throwing in a few program-specific buttons that help to place additional navigation and system control functions at your fingertips (or mouse click, as the case might be). Often, an elapsed time and measure/beat indicator is placed within this window, although certain sequencers may have a separate time/measure display window that can be easily resized, so that you can even keep track of your position in a sequence from across the room.

The track editing window is used to display such track information as the existence of track data, track names, MIDI port assignments for each track, program change assignments, volume controller values, and other transport commands. Depending on the sequencer, the existence of MIDI data on a particular track at a particular measure point (or over a range of measures) is indicated by the highlighting of that track range in a way that's highly visible. For example, in Figure 5.1, you'll notice that many of the measure boxes are highlighted. This means that these measures contain MIDI messages, while the other nonhighlighted measures don't. Other sequencers might highlight the boxes in different colors or have large dots at these measures to let you know where data exists.

Port assignments are often made from the track window. Often, track assignment boxes include a pop-up window that lets you graphically assign the port and MIDI channel for that track. In both the Mac and Windows

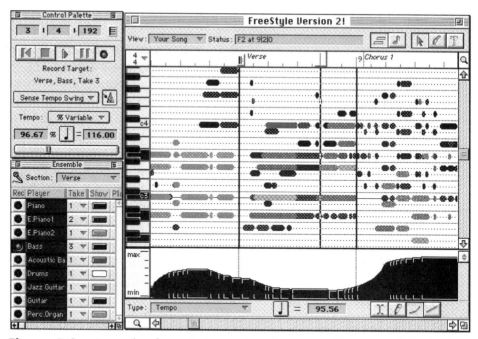

Figure 5.3 Example of FreeStyle's main edit screen. (Courtesy of Mark of the Unicorn, Inc., www.motu.com)

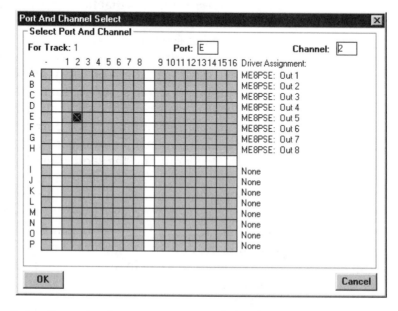

Figure 5.4 Example of a port assignment dialog box. (Courtesy of Passport Designs, Inc., www.passportdesigns.com)

environment, the number of ports available will depend on your interface type. For example, Figure 5.4 shows an assignment window that can access 8 ports of a multiport MIDI interface. Given that there are 8 ports and 16 channels per port, this sequencer could output MIDI data on up to 128 discrete channels.

Many sequencers offer an *auto channelizing* function that can automatically route incoming MIDI data from a controller source directly to a selected instrument. Basically, once a track is selected for recording and its ports have been assigned, the sequencer's auto channelizing feature will automatically route the incoming controller data directly out to that instrument's designated port and channel. In short, once an instrument has been assigned to a port and channel, all you need to do is click on its track record button and begin playing. The sequencer will take care of all the signal routing and channelizing for you.

Do It Yourself Tutorial: Auto Channelizing

1. Assign a few tracks to several voices or instruments, using different ports and/or MIDI channels.
2. Press the record button on one track. Does it automatically assign the controller to that port/channel?
3. Press the record button on another track. Did it automatically assign the controller to the new port/channel?
4. If so, your sequencer is auto channelizing the incoming MIDI to the selected track.
5. If not, check your manual to see if this feature is supported.

Another type of automation that's supported by most sequencing packages involves the use of *program change messages*. As we saw in Chapter 2, the transmission of a program change message (which ranges from 0 to 127) on a specific MIDI channel can be used to change the program or patch a preset number of an instrument or voice. By assigning a program change number to a specific sequence track, it becomes possible for an instrument (or a single voice within a polyphonic instrument) to be automatically set to the desired patch setting (Figure 5.5). In short, you can automatically assign an instrument or device patch to each track in a sequence, save and close the file, and then at a later time, when you open the file again, it will automatically recall all of your instrument patches. All you need to do is press the play button.

Note also that program changes can also be inserted in the middle of a sequence. Such a message can be used to instruct a synth voice to change from one patch to another in the middle of a song. For example, a synth could be used to play a B3 organ for the majority of a song; however, during

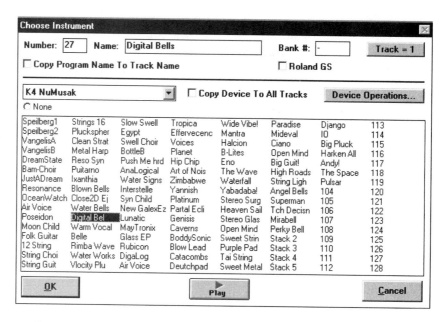

Figure 5.5 Program change messages can be used to assign patch settings automatically to a sequenced track.

a break, a program change message could instruct it to switch to an acoustic piano voice and then back again to finish the song. Patch changes such as these can also be used to change the settings for effects devices that can respond to MIDI program changes (i.e., changing a processor's settings from a rich room plate to a wacky echo setting in the middle of a sequence).

Although program change commands will often configure patches in a simple MIDI setup without much difficulty, more complex systems might not be so straightforward. Some heavy-duty manual reading and experimentation may be needed to fully automate your system's setup routine. For example, a synth that has literally hundreds of sound patches might require you to assign your favorite patches manually to one of the 128 possible program change numbers. Samplers can likewise have their own particular quirks for automatically recalling a sample or bank of samples from disk.

In addition to basic track and function assignments, most sequencers will let you vary continuous controller messages from the main edit screen. The most common and logical controller for quick access is the *volume controller* for an instrument or one of its voices. In short, this control is used as a volume control for changing the overall level of an instrument's voice. Although this is generally the default controller, certain sequencers will let you choose from any of the 128 possible controller types (see Figure 2.21).

For obvious reasons, standard tape transport track commands (such as record ready, solo, and mute) are often placed at the left-hand side of a track's control line. Most sequencers will also add a "loop" button, which can be used to repeat continually a short percussion or musical phase throughout an entire song.

The time signature/tempo window often simply lets you know what the current time signature (i.e., 4/4, 3/4, 4/5, etc.) and tempo (a song's metronomic speed) is. Often, more detailed time/tempo values can be entered into a score from a special dialog box.

"Chunks"

In addition to dealing with MIDI tracks in a linear fashion that flows continuously from the beginning to the end of a song on tracks 1, 2, 3, and so on, some sequencers work in a more free-form manner that lets you also group musical phrases into a defined "chunk" (Figure 5.6) that can be looped, moved, and processed as a single entity with extreme ease. This versatile type of sequencing works very well with musicians that like to assemble a song from loops and short musical phrases.

Piano Roll Editor

One of the easiest ways to see and edit MIDI track note values is through the use of the sequencer's *piano roll* edit window (Figure 5.7). This intuitive window lets you graphically edit MIDI data at the individual note level by displaying values on a continuous piano roll grid that is broken down into time (measures) and pitch (notes on a keyboard). By simply clicking on a note (whose beginning point and length is often displayed as a highlighted, horizontal bar), you can easily change it to a new value by vertically dragging it up or down the musical scale (often a keyboard is graphically represented at the window's left-hand side as a guide). It's often a simple

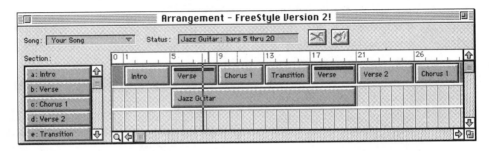

Figure 5.6 Region blocks can be used with some sequencers to assemble chunks into a finished song. (Courtesy of Opcode Systems Inc., www.opcode.com)

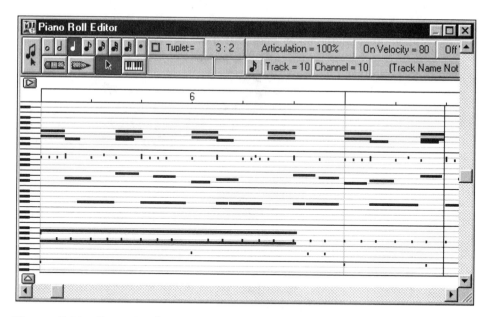

Figure 5.7 Example of a piano roll edit window. (Courtesy of Passport Designs, Inc., www.passportdesigns.com)

(A) Dragging a note vertically will change its pitch.

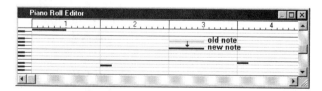

(B) Dragging the note horizontally will change its beginning start time.

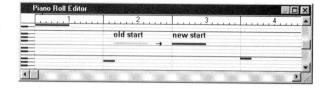

(C) Dragging the tail of a note will change a note's duration time.

Figure 5.8 Basic note editing in a piano roll edit window.

matter to drag the note horizontally across the window's time line, so that the note will begin at a different time. In addition, a note's duration can often be changed by clicking on the note's trailing edge and dragging the note out to the desired length (Figure 5.8).

As you might expect, the way in which notes can be edited in the piano roll window often changes from one sequencer to the next. The examples listed above are fairly common; however, certain sequencers might work in a different way or let you enter time and velocity as numeric values.

Notation Editor

Certain sequencing packages will also let you enter and edit notes into a sequence using standard musical notation (Figure 5.9). This window works in much the same way as the piano roll editor, except that pitch, duration, and other values are displayed and edited directly into measures as musical notes on the bass and/or treble clef.

One really nice by-product of this feature is that most sequencers that have this window will also let you print a track's musical part in standard notation. This can be really useful for creating basic lead sheets (lyrics can even be entered into some sequences) for a particular instrument or part. Only rarely do sequencers print all of the tracks in a sequence in a score form (where all the parts are arranged and printed onto the pages). Because the layout and printing demands are greater when laying out a printed

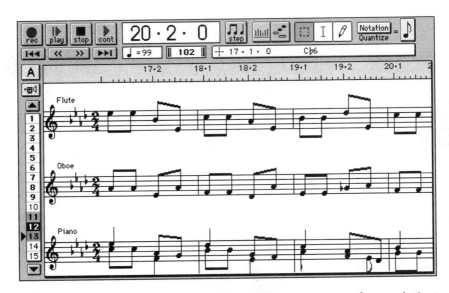

Figure 5.9 Musicshop's notation editing window. (Courtesy of Opcode Systems Inc., www.opcode.com)

score, most sequencer manufacturers offer a separate program for carrying out this fairly specific task.

Event Editor

Although they're used less often than either the piano roll or notation edit windows, an event editor lists all of the MIDI messages that exist on a track (or in the entire sequence) as a sequential list of events. For example, Figure 5.10 shows a lists of all MIDI events that occur in a MIDI sequence over time. At the measure point 7:2:045, we see that a note-on event is being transmitted for note D4 on channel 1 with a note-on velocity of 123 and a release velocity of 61.

This edit interface is often best for instructing the sequencer to perform an event (i.e., "Do something at such-and-such a time"), instead of being used as a musical editor. For example, you could instruct a PC-based sequencer to perform a media control interface (MCI) command to play a video clip (AVI file) or waveform (WAV file) at a specific time, or you could easily use this window to insert a program change on a certain channel that tells an instrument to switch voices on cue. Just a quick warning: For those of you who just got excited about the prospect of triggering a video or audio file directly from your sequencer, you should know that the start times (and therefore basic sync) are fairly unreliable. I urge you to read your sequencer's manual about imbedding MCI commands into a sequence and try it out for yourself.

Trk	Hr:Mn:Sc:Fr	Meas:Beat:Tick	Chn	Kind	Values		
3	00:00:12:16	7:2:044	1	Note	G 4	93	40
3	00:00:12:17	7:2:045	1	Note	D 4	123	61
3	00:00:12:17	7:2:046	1	Note	G 3	127	31
3	00:00:12:17	7:2:048	1	Note	E 4	103	11
3	00:00:12:23	7:3:000	1	Note	A 3	127	51
3	00:00:12:23	7:3:000	1	Note	E 4	127	63
3	00:00:12:23	7:3:000	1	Note	A 4	127	56
3	00:00:12:23	7:3:003	1	Note	G 4	123	10
3	00:00:13:15	8:1:004	1	Note	A 4	127	1:029
3	00:00:13:15	8:1:004	1	Note	E 4	127	1:059
3	00:00:13:15	8:1:004	1	Note	A 3	127	1:026
3	00:00:14:02	8:2:049	1	Note	G 4	100	36
3	00:00:14:02	8:2:050	1	Note	D 4	127	42
3	00:00:14:02	8:2:050	1	Note	G 3	127	27
3	00:00:14:08	8:3:003	1	Note	D 4	100	9

Event List - Track 3: Grunge guitar

Figure 5.10   Example of an event window. (Courtesy of Cakewalk, www.cakewalk.com)

Editing Controller Values

Most sequencers will let you enter or change controller message values from a simple, graphic window that allows you to draw a curve which graphically represents the effect that the controller will have on an instrument or voice (Figure 5.11). By using a mouse or other input device, it becomes a simple matter to draw a continuous stream of controller values that correspondingly changes such variables as velocity, modulation, pan, and other controllers. Changing a range of controller values is as easy as clicking the mouse and redrawing a new curve over the affected range.

Often, you'll find that these drawn curves have their resolution set so high that literally hundreds of controller changes can be introduced into the sequence over just a few measures. In most cases, this resolution simply isn't necessary and might even create a data bottleneck when playing back a complex sequence. To filter out the gazillions of unnecessary messages that could be placed into a track, you might want to lower the control change resolution (if your sequencer has such a feature) or you might want to thin the controller data down to a resolution that's more reasonable for the task at hand.

Do It Yourself Tutorial: Thin Controller Data

1. Open or record a MIDI track and select a few measures to be edited.
2. Open the edit controller window (or whatever name your sequencer uses).
3. Choose an obvious controller (i.e., #7, volume, or #1, modulation) and draw a curve.
4. Listen to the affected track.
5. Highlight the drawn curve and select the "thin controller data" function (see your manual or help file for info).

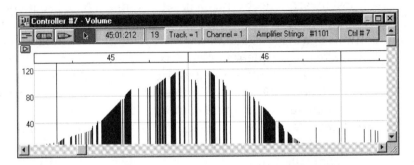

Figure 5.11 Example of a controller window. (Courtesy of Passport Designs, Inc., www.passportdesigns.com)

91

6. Did it remove tons of redundant controller messages?

7. Listen to the affected track again.

In addition to being able to draw and redraw controller changes, most sequencers will let you highlight a range of MIDI events on one or more tracks, and then process controller changes from the main edit window. This is really useful for selecting basic functions such as "strip controller data" (which effectively removes all or certain controllers), thinning controller data, and processing data (so as to change any note or controller variables).

Processing controller data can be useful for scaling controller data upward or downward. For example, you could use this feature to limit a track's overall dynamic range (without using a compressor). This can be done by scaling a track that might have velocity messages which range from 0 to 127, down to a range that varies from 64 to 96. You can process a range of messages so that they ramp up or down over a defined range, effectively creating a smooth fade-in or fade-out (something that's not always easy to do manually with a fader).

Do It Yourself Tutorial: Fade In a MIDI Track

1. Pick a sequence that has a basic intro track.

2. In the edit screen, select a range on a track (or tracks) that you'd like to fade in.

3. Select the window or function that can scale continuous controller #7 for that track range.

4. Tell the software to change velocity scaling from 0% to 100% (or from 0 to the track's original volume level).

5. Listen to your newly created fade!

Mixing a Sequence Using Continuous Controllers

Almost all sequencer types will let you mix a MIDI sequence. This is done by creating a physical or software interface that can emulate the physical controls of a basic audio mixer. Instead of directly mixing the audio signals that make up a sequence, these mixers are able to directly access such track controllers as main volume (controller #7), panning (controller #10), and balance (controller #8). In Figures 5.12 and 5.13, two types of sequencing software mixers are shown. Although their interface designs are different, they both offer a mixing surface that's easy to understand and operate.

We should also point out that since the mix-related data is simply MIDI controller messages, an entire mix can be easily stored within a sequence file. Therefore, even with the most basic sequencer, you'll be able to mix and remix your sequences with complete automation and total

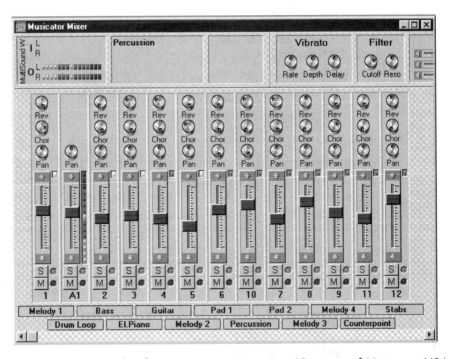

Figure 5.12 Example of a sequencer mix screen. (Courtesy of Musicator USA, www.musicator.com)

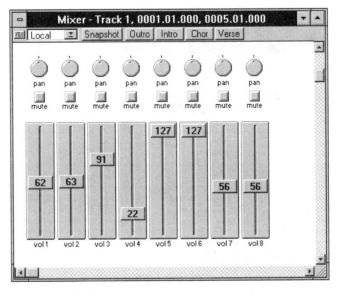

Figure 5.13 Example of a sequencer mix screen. (Courtesy or Steinberg N. America, www.steinberg-na.com)

settings recall, whenever a new sequence is opened. As an added bonus, it's quite common for the mixer interface to have "moving faders and pots." This means that moves that have been created during a mix will automatically update and change on the mixer screen right before your eyes. This feature is not only useful for letting you know where your settings are, it's also fun!

Changing Tempo

One of the advantages of recording into and playing back a musical sequence is the ability to change the tempo of a song easily without changing the program's pitch or real-time control parameters. Basically, all you need to do is change the tempo of a sequence (or part of a sequence) to best match the feel of the song, or to slow the speed temporarily so that you can more easily play along with previously recorded tracks.

Some sequencers can be programmed to change tempo automatically within a song, using a feature called *tempo mapping* (Figure 5.14). Often tempos can be visually "mapped" by a software sequencer through the use of a window display that lets you change the meter and possibly the time signature over the course of a song.

Some sequencers will offer a "fit time" feature that can automatically adjust a song's tempo mapping so as to stretch or shrink it to a specified time length. For example, this function could easily be used to

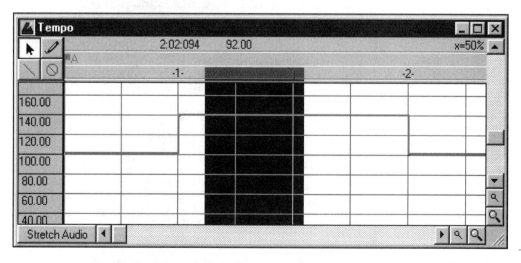

Figure 5.14 Example of a tempo map window. (Courtesy of Cakewalk, www.cakewalk.com)

micro-tune the tempo of a 37-second jingle to snugly fit into a 30-second time slot.

Practical Editing Techniques

When it comes to learning the ins, outs, and thrus of basic sequencing, absolutely nothing can take the place of diving in and experimenting with your setup. I'll paraphrase a buddy of mine, Craig Anderton, who said in his book *Power Sequencing with Master Tracks Pro/Pro4:* "Read the manual through once when you get the program (or device), then play with the software and get to know it *before* you need it. Afterwards, reread the manual to pick up the system's finer operating points." Wise words!

In this short section, we cover some of the basic techniques that will speed you on your way to sequencing your own music. Note that there are no rules to sequencing MIDI. As with all of music (and the arts, for that matter), there are as many "right" ways to perform and sequence music as there are electronic musicians. For example, I often work in a slower, free-form musical style that works best when sequenced without a click track. Others that work in more traditional mainstream meters wouldn't think of working without this metronomic guideline. Sequencing also lends itself well to basic kamikaze approaches to making music. For example, I've been known to take MIDI files off the Internet, cut and paste the heck out of them, slow them down, and then assign them to voices that are totally different than were originally intended.

For the remainder of this section, let's take a look at some general sequencing techniques that can help you get started. Just remember that these are but a few of the possible guidelines and tips. After a while, you'll discover your own guidelines and working procedures. Basically, I'm here to tell you that the field's wide open, and that there's no right or wrong. Just experiment, have fun, and make music!

Transposition

Changing the pitch of a note or the entire key of a song is extremely easy to do with a sequencer. Depending on the system, a song can be transposed up or down in pitch at the global level, thereby affecting the musical "key" of a song. A segment can be shifted in pitch from the main edit window by simply highlighting the bars and tracks that are to be changed, and then calling up the transpose function from the system's menu. Transpositions can likewise be made from either the piano roll or the notation editor. This is done by highlighting the notes to be changed and then transposing them according to the system's manual.

Quantization

By far, most common timing errors begin with the performer. Fortunately, "to err is human," and standard performance timing errors often give a piece a live and natural feel. However, for those times when timing goes beyond the bounds of nature, a sequencing feature known as *quantization* can help correct these timing errors.

Quantization lets you adjust timing inaccuracies to the nearest desired musical time division (such as a quarter, eighth, or sixteenth note). For example, when performing a passage that requires that all involved notes fall exactly on the quarter-note beat, it's often easy to make timing mistakes (even on a good day). Once the track has been recorded, the sequencer can recalculate each note's start and stop times so that they fall precisely on the boundary of the closest time division (Figure 5.15). Such quantization resolutions often range from full whole-note to sixty-fourth-note values.

Because quantization is used to "round off" the timing elements of a range of notes to the nearest selected beat resolution, you are advised to try and "lock" your playing in time with the sequencer's own timing resolution. This is simply done by selecting the tempo and time signature that you want and turning on the click track. By playing along with a click track, you're basically using the sequencer as a musical metronome. Once the track or song has been sequenced, the quantization function can be called up, which will further correct the selected note timing values.

If you were to record the same sequence without an initial timing reference (most often a click track), the notes would be quantized to a timing benchmark that doesn't exist. That's not to say that it is impossible to quantize a sequence that wasn't played to a click—it's just trickier business. In

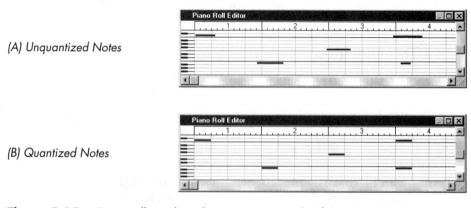

(A) Unquantized Notes

(B) Quantized Notes

Figure 5.15   Piano roll window showing an example of the quantization process to a quarter-note resolution. (Courtesy of Passport Designs, Inc., www.passportdesigns.com)

short, whenever you feel you might need to quantize a segment, always give special consideration to a sequence's initial timing elements (i.e., the sequencer's click track).

Humanizing

When you get right down to it, one of the most magical aspects of music is its ability to express emotion. A major factor used for conveying expression is the minute timing variations that are introduced during a performance. Whenever these variations become so large that they become sloppy, the first task at hand is to tighten up the section's timing through quantization. The downfall of quantization, however, is that its "robotic" accuracy can take away from the basic human variations in the music, making it sound rigid and machine-like. One of the ways to reintroduce these variations in timing back into a quantized segment is through a process known as *humanization*.

The humanization process is used to randomly alter all of the notes in a selected segment according to such parameters as timing, velocity, and note duration. The amount of randomization can often be limited to a user-specified value or percentage range, and parameters and can be individually selected and/or fine tuned for greater control (Figure 5.16).

Beyond the obvious advantages of reintroducing human-like timing variations back into a track, this process can help add expression by randomizing the velocity values of a track or selected tracks. For example, humanizing the velocity values of a percussion track that has a limited dynamic range can help bring it to life. The same type of life and human "swing" can be effective on almost any type of instrument. Let's give it a try. . . .

Do It Yourself Tutorial: Quantize and Humanize a Track

1. Pick a sequenced track (preferably a percussion track) that has a lot of rhythmic activity.

Figure 5.16
Example of a humanize dialog box. (Courtesy of Passport Designs, Inc., www.passportdesigns.com)

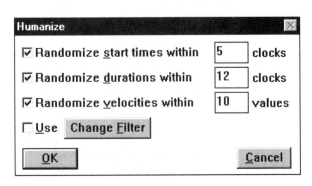

2. Select a range of measures and play with the quantize feature until you're happy with the results.
3. Call up the humanize function and experiment with various time, velocity, and length variables. (You can almost always undo your last move, before trying a new set of values.)
4. Listen to your newly humanized track!

Slip Time

Another timing variable that can be introduced into a sequence to help change the overall "feel" of a track is the *slip time* feature (Figure 5.17). Slip time is used to move a selected range of notes either forward or backward in time by a defined number of clock pulses. This has the obvious effect of changing the start times for these notes, relative to the other notes or timing elements in a sequence.

The slip time function can be used to micro-tune the start times of a track, so as to give them a distinctive feel. For example, nudging the notes of a sequenced percussion track forward in time by only a small number of clock pulses will effectively make the percussion track rush the beat, giving it a heightened sense of urgency or expectation. Likewise, retarding a track by any factor will give it a slower, backbeat kind of feel.

Slipping can also be used to move a segment manually into relative sync with other events in a sequence. For example, let's say that we've created a song that grooves along at 96 bpm, and that we've searched our personal archives and found a bridge (a musical break motif) that would work great in the piece. After inserting the required number of empty measures into the sequence and pasting the break into it, we found that the break comes in 96 clocks too late. No problem! We can simply highlight the break and slip it forward in time by 96 clocks. Often this process will take several tries and nudging to find the timing that feels right, but the persistence will definitely pay off.

Figure 5.17
Example of a slip time dialog box. (Courtesy of Passport Designs, Inc., www.passportdesigns.com)

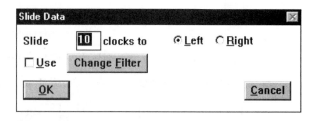

Work Those Tracks!

For decades, I've had what I call Huber's law: There are no rules, only guide-lines! Well, this wise adage also applies to the creative process of manipu-lating, combining, and playing with MIDI tracks to create that all-powerful musical experience. A whole host of editing and creative tools are at your disposal to breathe more life into a sequenced track or set of tracks; a few of these are track splitting, track merging, voice layering, and echoing.

Splitting Tracks

Although a track is encoded as a single entity that includes musical notes and other performance-related events, the data that's contained in a track can often be separated for further processing or routing to another instru-ment. As an example, let's say that we're working on a sequence that in-cludes a synth part that was played live on stage. You, as the producer, have decided that you'd like to have a greater degree of level mixing control over the left-hand part, while leaving the bass part alone. This isn't really a prob-lem. The best solution would be to split the track into two sequenced tracks. This is done by making an exact copy of the synth track and pasting it onto a separate track. Once done, you can call up the piano roll window for the original track (so you can see and graphically edit the note ranges), highlight the bass notes, and delete them. Then you can repeat the process on the copied track, deleting the upper notes. By assigning the two tracks to two separate voices or by altering the velocity values of each track and assign-ing it to the same voice, you can now have a greater degree of control over the parts.

Merging Tracks

Just as a track can be split into two or more tracks, multiple tracks can be *merged* into a single track. On most sequencers, this can be done easily by highlighting the entire track or segment that you want to merge and copy-ing it into the sequencer's memory. Once done, you can select the destina-tion track and measure into which the copied data is to be merged and then invoke the sequencer's merge command.

When merging MIDI data, you should keep in mind that it might be difficult to unmerge data that has been combined into a single track. For this reason, when using a sequencer that has plenty of tracks to spare, you might want to save the original unmerged tracks and simply mute them. They won't be played back and they'll be totally intact should you need them later.

Another use of track merging involves the playing back of an instrument track while recording real-time controller data from a pitch, modulation, or other type of controller onto a separate track. This is done by recording the controller data onto an open track that's assigned to the same MIDI channel as the instrument track and then adjusting the controller while listening to the instrument. I often find that this improves a performance, because it frees you up to concentrate on your performance and lets you deal with controller changes at a later time. Once the controller track has been made to your liking, you can merge it with the original track (or keep them separate for future editing or archive purposes).

Layering Tracks

One of the more common and powerful tricks of the trade that's used by seasoned electronic musicians is *layering* instrument voices together to create a new, composite sound. Although digital samples and modern-day synthesis techniques have improved over the years, you're probably aware that many of these sounds still have a character that can be easily distinguished from their acoustic counterparts.

One way to fill out a sound and make it richer and more realistic is to layer instruments together into a single complex voice. One example of layering would be to take the sound of a piano from synthesizer A and combine it with the sounds of a sampled Steinway from sampler B. This process can be done by making a copy of the original track and then assigning these tracks to their respective ports, channels, and instrument voices.

Often layers can be made to sound even more realistically complex by humanizing each track. The variations in velocity, beginning, and length times will simply help to make the sounds more convincing. Of course, because the tracks are separate, they can be combined in the mix in ways that will achieve the best and/or most convincing results.

MIDI Echo, Echo, Echo . . .

They say that you can never have too many effects boxes in your toy chest. Well, MIDI can also come to the rescue to help you set up effects in more ways than you might expect. For example, a voice can be easily repeated in a digital delay fashion by copying a track to another track and then slipping that track (either forward or backward) in time. By assigning these tracks to the same destination (or by merging these tracks into one), you'll be setting up a fast and free MIDI echo effect. Simply repeat the process if you want to add more echos.

SysEx

As you may have guessed from comments in previous chapters, I'm a big fan of the power and versatility that System Exclusive (SysEx) can bring to almost every electronic musician's project studio. Most sequencers are able to read and transmit this instrument and device exclusive data, allowing you to create your own sounds, grab sound patches from the Internet, swap patches with your friends, or buy patch data disks from commercial vendors.

In summary, SysEx data is used directly to access, communicate, and store a synthesizer's internal patch and setup data. However, it can also be used to access the memory presets of most MIDI devices (such as effects devices, MIDI patchbays, or the internal setup and rhythm patterns of a drum machine). All of this can be done by communicating SysEx data directly between MIDI devices of the same type, or it can be done by sending the data to a sequencer or sequencer software package, saving it to disk as a named file and then transmitting it back to the device at a later time.

Another use for SysEx that many folks overlook is its ability to act as a backup medium for storing your patch and setup data just in case your system's memory gets corrupted or your instrument's RAM memory battery decides to go belly up. Here's an example of how a SysEx backup can save the day. A few years ago, I crashed the data in my WaveStation SR's memory by loading in a patch from a WaveStation EX that wasn't totally compatible. Everything worked until I got to a certain patch and then the box went into a continual reboot loop—major freak-out time! No matter what I did, the system was totally locked up! The solution? I opened up the box and took out the backup memory battery (which in turn cleared out the error, as well as the box's RAM patch memory). All I had to do was reload the factory SysEx patch settings into the box and I was back in business in no time.

Oh, yeah, I almost forgot. I've also run into situations where SysEx data has gotten corrupted over the course of a few years as it passed through an instrument, was modified (we all love to tweak our setups from time to time), and then saved as a new file. For this reason, I've found it wise to keep an early generation set of SysEx backups on the shelf for such an unlikely event. After all, an outdated backup is better than a lost set of patches.

Backup? Did someone say "backup?" Speaking of backup, this one much-overlooked word can mean the difference between a saved day and relative disaster. One of the nice things about MIDI data is that the file sizes are small. They can easily be saved to a floppy, hard disk, or backed up en masse to recordable CD media. When in doubt, back up your files. When not in doubt, back up your files. You'll thank yourself for it.

Playback

Once a sequence is composed and saved to disk, all of the sequence tracks can be transmitted through the various MIDI ports and channels to the instruments or devices to make beautiful music, create sound effects for film tracks, or even control device parameters in real time.

Because MIDI data exists as encoded real-time control commands, you can listen to the sequence and make changes at any time. You could change the patch voices, alter the final mix, or even change and experiment with such controllers as pitch bend, modulation, or after touch—even change the tempo and key signature. In short, this medium is infinitely flexible in the number of versions that can be created, saved, folded, spindled, and even mutilated until you've arrived at the overall sound and feel that you want. Once done, you'll have the option of using the data for live performance and/or mixing the tracks down to a final recorded media (such as DAT or hard disk, and ultimately to CD), either in the studio or at home.

During the summer, in a wonderful small-town tavern where I live, there's often a performer that'll wail the night away with his voice, trusty guitar, and a backup band that consists of several electronic synth modules and a laptop PC/sequencer that is just chock-full of Country-n'-Western sequences. His set of songs for the night is loaded into the *song playlist* feature that's programmed into his sequencer. Using this playlist, the sequences are sequentially queued so that when one song finishes and he has taken his bow, introduced the next song, and complimented the lady in the red dress, all he needs to do is press the space bar and begin playing the next song. Such is the life of an on-the-road sequencer.

In the studio, it has become more the rule than the exception that MIDI tracks will often be recorded and played back in sync with either an analog multitrack machine, digital multitrack, or hard disk recording system. Whenever more than one playback medium is involved in a production, a process known as *synchronization* (Figure 5.18) is required to make sure that events in the MIDI, analog, digital, and even video media occur at the same point in time. The specifics of how events can be "locked" together in time are accomplished in various ways (depending on the equipment and media type used); however, the most commonly used coding language used throughout the world in production today is SMPTE time code (short for Society of Motion Picture & Television Engineers).

The use of time code (its more common name) in a synchronized system makes it possible to link MIDI tracks, multitrack recorders, hard disk tracks, and visual media together, such that when a sequence or other media is played at any point within the song, all of the other devices will locate to

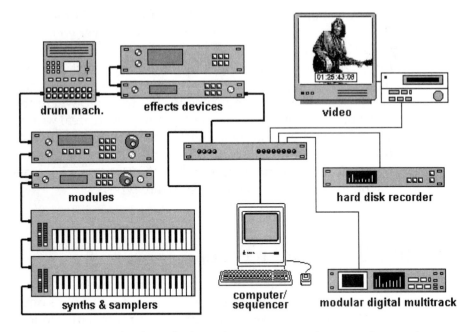

Figure 5.18 Example of a multiple-media music system that is synchronized in time via SMPTE time code.

that point and begin to play in synchronous time. More in-depth reading on synchronization can be found in Chapter 10.

Standard MIDI Files

Before ending our discussion on sequencers, you should be aware that there are as many proprietary file formats for saving MIDI- and sequence-related projects to disk as there are sequencing manufacturers. As a result of the relative chaos of trying to import a sequence from manufacturer A into one made by manufacturer B, a standard was developed that allows sequence files to be freely interchanged between different programs using the same or different computer systems. This format (known as *standard MIDI files*) can encode time-stamped MIDI event data, as well as song, track, time signature, and tempo information. Standard MIDI files are able to support both single and multichannel sequence data, and are commonly entered into the system through the sequencer's load/save or import/export functions. These file types, which are stored using the .MID extension, are also the format of choice for use on the Internet and for distributing MIDI files to the general masses.

You should keep in mind that standard MIDI files come in two basic flavors: type 0 and type 1. Type 0 is used whenever you would like all of the tracks in a MIDI sequence to be compressed into a single MIDI track. All of the original channel messages would remain intact, but the data wouldn't have definitive track assignments. This data type might be the best choice when creating a MIDI sequence for the Internet, where the sequencer or MIDI player application might not know how to deal with multiple tracks. The type 1 format, on the other hand, will most likely retain its original track structure and can be imported into another sequencer type with its track information and assignments left intact (to varying degrees depending on the involved systems or software types).

Other Sequencer Types

In addition to standard music sequencing, a few other sequencer types are available that can perform specific tasks. These systems often relate to such specific types of music production as drum pattern sequencing and algorithmic music composition.

Drum Pattern Editor/Sequencers

At any one time, there is generally a handful (OK, maybe two handfuls) of companies that have software or hardware devices that's specifically designed to create and edit drum patterns. These systems often rely on user input and quantization to construct any number of percussion grooves; however, more often than not they rely on a grid pattern display that places MIDI notes representing the various drum sounds along the vertical axis, while time is represented in metric divisions along the horizontal axis (Figure 5.19). By clicking on each point in the grid with a mouse or other input system, individual drum or effect sounds can be built into patterns that are hopefully rhythmically diverse and interesting.

Once created, these and other patterns can be linked (the other term being *chained*) to create a partial or complete song. These editors commonly offer the ability to change MIDI note values (thereby changing drum voices), quantization, humanization, and adjust note and pattern velocities. Once completed, the sequenced drum track (or series of tracks) can be saved in the standard MIDI file format for import into a standard music sequencer.

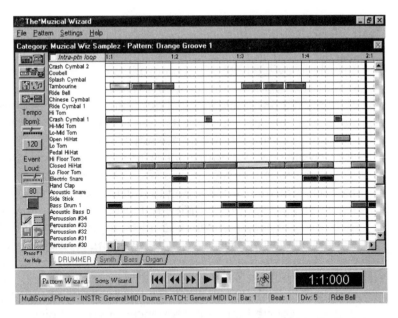

Figure 5.19 Example of a drum sequencer. (Courtesy of Media Tech Innovations, www.midibrainz.com)

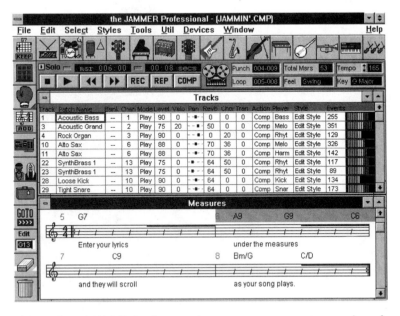

Figure 5.20 The JAMMER Professional interactive composing and performing program. (Courtesy of Soundtrek, www.soundtrek.com)

Algorithmic Composition Programs

Algorithmic composition programs (Figure 5.20) are interactive sequencers that directly interface with MIDI controllers or standard MIDI files to generate MIDI performance data according to a computer algorithm. In short, once you give it a few basic musical guidelines, it can begin to generate performances or musical parts on its own. This might help you gain new ideas for a song, create automatic accompaniment, make improvisational exercises, create special performances, or just plain have fun.

This sequencer type uses a wide range of program parameters to control the performance according to musical key, notes to be generated, basic order, chords, tempo, velocity, note density, rhythms, accents, etc. Often such interactive sequencers will allow input from multiple players, allowing it to be manipulated as a collective jam. A single player can also record and manipulate a composition or record several interactive compositional parts that can then be combined and manipulated at any time. Once a composition has been satisfactorily generated, a standard MIDI file can be created and exported into any sequencer of your choice.

6

EDITOR/LIBRARIANS

As we've seen from Chapter 4, the vast majority of MIDI instruments and devices store their internal patch data within RAM memory. Synthesizers, samplers, or other devices contain information on how the controlling oscillators, amplifiers, filters, tuning, and other presets are to be configured, in order to create a particular sound timbre or effect. In addition to controlling sound patch parameters, a unit's internal memory can also store such setup information as effects processor settings, keyboard splits, MIDI channel routing, and controller assignments. These parameters, which can be manually edited from the device's front panel, can be easily accessed by recalling the patch number and/or name via the device's bank of preset buttons, alpha dial, or keypad entry (Figure 6.1).

A device's initial preset settings can exist in several forms. The most basic of these is the factory preset. As the name implies, a factory preset is a bank of patches that have been programmed by the manufacturer to give the performer an initial set of sounds that are designed to be useful out-of-the-box and to *wow!* the potential customer into making a purchase. In almost every circumstance, these factory presets are encoded into battery-backed random access memory (RAM), where they will be stored

Figure 6.1
Example of a bank of preset buttons for storing and recalling patch data.

until either the battery loses power (generally over a 5-year period) or the patch data is overwritten. In some circumstances, these factory presets can be restored by reinitializing the synth, which is often done by pressing a hidden button or by pressing a specific button combination (check your manual to see if this feature is supported).

In the final analysis, however, factory patches are simply a collection of someone else's sounds and might not reflect your own taste and style. Many musicians, in fact, think that you're doing well if you like even half of the sounds that are supplied with an instrument. This alone is incentive enough to set even the most technologically timid on the road to editing and creating their own sound and parameter patches.

Sound- and setup-related patch data can be edited in a number of ways. One of the fastest and simplest ways is to alter the parameters of an already existing factory preset until a desired voicing effect or setup is achieved. Once done, you can overwrite the current patch setting or save the edited setting to a new patch location. This process of editing an existing patch is often easy to do, because half the programming work will have already been done for you in the original factory preset.

A second, more adventurous, approach is to build a patch entirely from scratch. In this way, a totally new sound can be constructed either by trial and error (by listening to the effect as the parameters are being changed) or by using your own experiences in synthesis and waveform analysis to build a patch from scratch (a process that's not for the faint-hearted). For example, a wavetable synthesizer patch setting might be programmed as follows:

- Experiment with playing the various raw (nonprocessed) wavetable sounds, until you find one that you like.
- Enter the edit mode, where you can process the way in which the PCM sample (or sample loop) is played back by adjusting such controls as frequency oscillators, amplifiers, filters, tuning, effects processor settings, keyboard splits, and controller assignments.
- If the unit is polyphonic and can assign and mix more than one voice on the same channel, you might want to mix in other sounds with the original and edit them accordingly until you have created a composite sound that you're proud of.

I know that I made the above process sound relatively simple, but it often isn't. Editing a patch from scratch can take a lot of time and can be very tedious. However, for those who have become good at it (all things come easier with practice), you have the satisfaction of showing your patches off in your work or, if you're really good at it, you might even consider becoming a professional programmer for an instrument manufacturer.

A Historical Perspective

Earlier analog synthesizers were constructed in a modular building block approach that made it easy to see and physically control knobs and/or sliders in a hands-on fashion. For example, an oscillator could be amplitude modulated by physically patching it (hence, the term *patch*) into a VCA (*voltage-controlled amplifier*) and then into a filtering module, etc. This process continued until the desired sound is achieved. The down side of this approach was that these controls required a large amount of space, were rather expensive, and made it difficult (if not impossible) to store settings for later recall. It's interesting to note, however, that this hasn't stopped the resurgent popularity of devices that have this hands-on type of control surface (from both a collector and new product standpoint).

With the advent of digitally controlled analog and fully digital synthesizers, most of the control surfaces have been redesigned for mass manufacturing, cost, and space-saving reasons. Newer display types have replaced the vast array of individual controls with a central control panel that's often made up of a liquid crystal display (LCD), buttons, alpha dials, and keypads. The down side of this modern-day approach is that you often have to spend much of your time navigating through the deep layers of programming options from the ever-tedious LCD panel (which limits your view of the techno-world to two lines on a 2-inch by 5-inch or smaller screen).

The Patch Editor

Because of this multilayered jungle, electronic musicians have taken back control of the control parameters of a MIDI device through the use of a computer-based patch editor. A *patch editor* is a software package that's used to provide on-screen controls and graphic windows for emulating hands-on controls that can often be varied in real time. A dedicated patch editor is device specific in that it has been programmed to emulate and vary the editing controls of a single instrument or device type. In certain cases, a patch editor might be designed to control several versions of an instrument (i.e., a Korg WaveStation, WaveStation EX, and WaveStation SR). However, because the coding is device specific, a patch editor for one instrument can't edit another instrument type.

Direct communication between the software/computer system and the device's microprocessor is most commonly accomplished via the real-time transmission and reception of MIDI SysEx messages (Figure 6.2). As noted in Chapter 2, SysEx messages are used to communicate customized

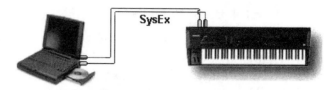

Figure 6.2 A simple example of SysEx data distribution between a patch editor and a MIDI instrument.

MIDI messages (such as patch parameter data) between MIDI devices. The format for transmitting these messages includes a SysEx status header, manufacturer's ID number, any number of SysEx data bytes, and an EOX byte. Through the transmission of these device-specific messages, the device's microprocessor can be directly accessed so that a large number of setup parameters can be directly and easily altered in real time. Note that Figure 6.2 shows a MIDI line that returns from the receiving device back to the computer. This was included to show that certain devices must be in constant two-way communication (often called a handshake protocol) with the editing software in order to work. (Again, check your device's manual to see if handshaking is required.)

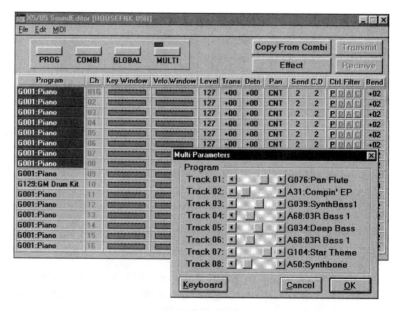

Figure 6.3 The Korg X5/05 SoundEditor. (Courtesy of Korg USA, Inc., www.korg.com)

The way that patch editors are graphically laid out on a computer monitor often varies widely from one software manufacturer to the next and may depend on the function and type of device parameters that are to be edited. For example, certain editors are designed to interact with a MIDI device strictly by using a numeric interface (this especially holds true for earlier shareware editors). Most newer and commercial editing software packages display device parameters and variables using a graphic display to manipulate the settings directly by simply placing a mouse cursor in the selection area and entering the new value or by dragging graphs to a new setting (Figure 6.3).

Almost all popular voice and setup editing packages include provisions for receiving and transmitting bulk patch data between the computer and MIDI device. This makes it possible to save and organize large amounts of patch data files to floppy or hard disks. Many patch editors can also print patch parameter settings or make a print of all of the patches within a particular bank or overall instrument banks.

Hardware Patch Editors

In addition to software editing packages, hardware solutions are available for gaining quick and easy access to device parameters via SysEx. Although they aren't common, certain synthesizer manufacturers offer device-specific hardware control surfaces that plug directly into the synth (Figure 6.4). For obvious reasons of speed, ease of operation, and hands-on flexibility, these edit panels have become popular with musicians that have equipment which supports this option.

Another option that's much less device specific is a MIDI data fader controller. These controllers, which are often equipped with 8 or 16 data faders, are commonly used to mix volume levels in the MIDI domain (see Chapter 11). However, these controllers can also be programmed to control

Figure 6.4
The LaunchPad MIDI controller. (Courtesy of E-mu Systems, Inc., www.emu.com)

directly the parameters of several different instruments or device types in real-time via SysEx.

Universal Patch Editor

In addition to the dedicated patch editor that is designed to work with a specific instrument, a number of programs are available that can communicate with a wide range of MIDI instruments and effects devices. These programs, known as *universal editors* (Figure 6.5), are designed to receive and transmit device-specific SysEx data and to provide on-screen control over the programming functions of most (if not every) MIDI device in your system.

Understandably, creating of a graphic representation of the various faders, knobs, and other parameter controls for every possible MIDI instrument is difficult. To do this, many universal editors are able to conform their interface type to the large instruments and devices that are on the market via a standard set of sliders, numeric input, and graphic windows. Some programs make the editing process a bit more intuitive by graphically representing (either loosely or with relative accuracy) the basic controls of certain selected devices. The deeper parameters will usually be edited using numeric or other graphic display types. A few graphically oriented universal editors will let you design your own customized controls for editing

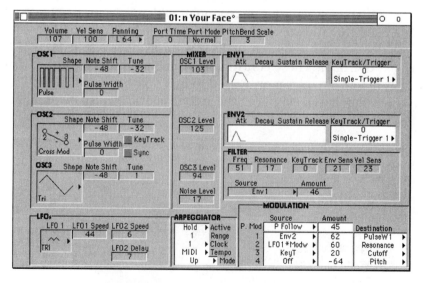

Figure 6.5 Waldorf Pulse edit screen of the Galaxy universal editor. (Courtesy of Opcode Systems Inc., www.opcode.com)

new instruments that have just reached the market or are yet to come, or even for editing those weird devices that no one would otherwise spend the time creating an interface for.

The Patch Librarian

After editing the original factory patches or after creating new patches from scratch, sooner or later, your instrument or device will invariably run out of preset memory locations. When this happens, hopefully you've learned that you can transmit these patches as SysEx data to either your sequencer's SysEx utility or another SysEx-related program. Once done, you can keep on creating new patches *ad infinitum*—and when you want a patch from a previous edit bank, simply reload that SysEx dump into the device and you're back in business.

After a while, however, having patches spread over tons of SysEx dumps might be too much of a good thing. You might end up spending more time loading patches than you'd like. In such a case, a software-based utility, known as a *patch librarian* (Figure 6.6), comes in handy. A patch librarian can be used to take the patches from various patch dumps and organize them into a single dump that contains similar patches. For example, you could use a patch librarian to organize all of your sustaining pads into

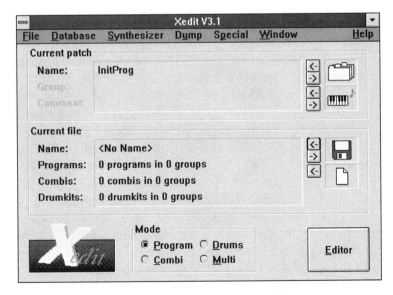

Figure 6.6 Xedit editor/librarian for Korg 05- and X-series synthesizers. (Courtesy of Joost Nieuwenhuijse, http://electron.et.tudelft/~joostn/korg.html)

a dump called "NewAge1" or all your hard hittin' patches into "HeavyMetal." You can also reorganize patches into banks according to sound type (i.e., bass, strings, effects) or any other way, for that matter.

Patch librarians come in two flavors: They may be dedicated to organizing and communicating patch data with a specific instrument or device or, more commonly, they can organize and communicate patch data with a wide range of MIDI instruments or devices (this latter type is commonly known as a *universal patch librarian*).

Sequencer packages that integrate a universal patch librarian into their program package can often automatically transmit appropriate patch data, setup, and program change messages to each device within a MIDI system. In this way, the librarian can be used to reconfigure the voice and setup patches throughout a system in advance of playing the sequence.

Alternative Sources for Obtaining Patch Data

In addition to factory- and user-created custom patches, many other sources exist where you can obtain patch data from both commercial and public sources. Preprogrammed patches are available for almost every popular MIDI instrument and effects device, spanning almost every timbre and instrumentation style. These patch banks have been programmed by working professionals and dedicated gearheads using a wide range of formats, including computer disks, ROM or RAM cards, and CD-ROMs. These professional and homegrown products can be commonly found at a reasonable cost in the classified section and back pages of most magazines that cater to the electronic musician.

ROM and RAM cards, which physically plug into supporting electronic instruments, can give you access to new patches (and in some cases, additional PCM wavetable samples). These credit card-like memory chips are often available from the instrument manufacturers or from third-party vendors that can be found in the classified section and back pages of most magazines that cater to electronic musicians.

Last, but definitely not least, is the Internet. Surfing the web will probably give you instant access to more patch data than you might know what to do with. An easy way to get started is to log on, go to your favorite search engine (such as yahoo.com or altavista.com), and simply type in the name of the instrument for which you want to search.

Most likely, the SysEx data on the Internet will be encoded using an application that is standardized by the general pubic at large (midiex.exe for the PC is one that comes to mind). The need to use such an application stems from the nonstandardization of the SysEx file format between sequencer and SysEx utility programs. All you need to do is download the utility that was used to encode the downloaded SysEx data and use the program to transmit the data to your instrument or device. Once this is done, you can transmit the patch data from the instrument back to your sequencer's SysEx utility, where it can be saved and retransmitted throughout your system with relative ease.

MUSIC PRINTING PROGRAMS

In recent years, the field of transcribing musical scores onto paper has been strongly affected by both the computer and MIDI technology. This process has been enhanced through the use of newer generations of computer software that makes it possible for music notation data to be entered into a computer either manually (by placing the notes onto the screen via keyboard and/or mouse movements) or by direct MIDI input. Once entered, these notes can be edited in an on-screen environment that lets you change and configure a musical score or lead sheet using standard cut-and-paste editing techniques. In addition, most programs can play the various instruments in a MIDI system directly from the score. A final and important program feature is their ability to print hard copies of a score or lead sheets in a wide number of print formats and styles.

Entering Music Data

A *music printing program* (also known as a *music notation program*) will let you enter musical data into a computerized score in a number of manual and automated ways (often with varying degrees of complexity and ease). Programs of this type offer a wide range of notation symbols and type styles that can be entered either from a computer keyboard or mouse. In addition to entering a score manually, most music transcription programs will generally

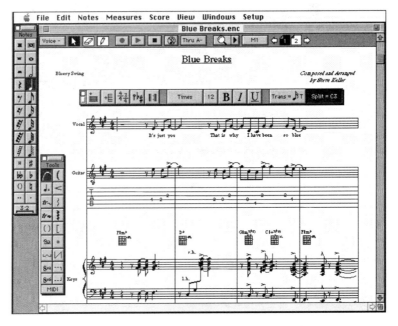

Figure 7.1 Encore music printing program. (Courtesy of Passport Designs Inc., www.passportdesigns.com)

accept MIDI input, allowing a part to be played directly into a score. This can be done in real time (by playing a MIDI instrument or finished sequence into the program), in step time (entering the notes of a score one note at a time from a MIDI controller), or from a standard MIDI file (which uses a computer sequence file as the source).

Many manufacturers of professional sequencing software will also offer an introductory or professional music printing program, that will almost always let you enter a completed sequence directly into the notation software using the sequencer's native file structure. This often helps to translate information such as titling and a basic time signature that would otherwise be lost when importing a standard MIDI file.

Music printing programs (Figure 7.1) often vary widely in their capabilities, speed of operation, and number of offered features. These differences include such features as the number of musical staves (the lines that music notes are placed onto) which can be entered into a single page or overall score, how many parts can be placed into a score, the overall selection of musical symbols, the ability to enter text or lyrics into a score, etc.

Certain software packages limit notation to a single grand staff (which is composed of an upper treble clef and lower bass clef), while

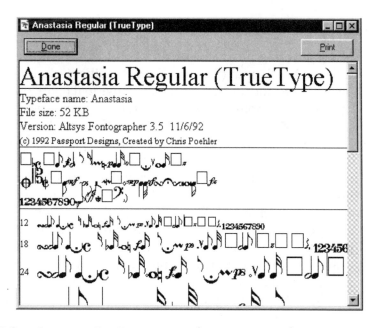

Figure 7.2 Anastasia TrueType notation font. (Courtesy of Passport Designs Inc., www.passportdesigns.com)

other more comprehensive programs let you enter notes onto a larger number of staves for creating larger compositions and orchestral scores. A wide selection of musical symbols is commonly available and can be placed into a score to denote standard note lengths, rest duration markings, accidental markings (flat, sharp, and natural), dynamic markings (i.e., pp, mp, mf, and ff), and a host of other important score markings (Figure 7.2).

In addition to standard note entry, text can be entered into most lead sheets or scores for the purpose of adding song lyrics, song titles, special performance markings, and additional header/footer information. Text can often be easily edited, allowing lyrics to be cut, copied, and pasted.

Scanning a Score

Another way to enter music into a score is through the use of an optical recognition program. These programs let you place sheet music or a printed score into a standard flatbed scanner, scan the music into a program, and then save the notes and general layout as a NIFF notation format file. This newly created standardized file format can be directly read by most professional printing programs for further editing, printing, and playback.

Editing a Score

Music printing programs often let you edit music manually or automatically via MIDI input (Figure 7.3). Using either technique it's generally a simple matter to add, delete, or change individual notes, duration, and markings by using a combination of computer keyboard commands, mouse movements, or MIDI keyboard commands. Of course, larger blocks of music data can also be edited using standard cut-and-paste methods.

One of the major drawbacks to entering a score via MIDI (either as a real-time performance or from a standard MIDI file) is the fact that music notation is an interpretive art. "To err is human" and this human feel is what usually gives music its full range of expression. It's very difficult, however, for a program to properly interpret these minute, yet important imperfections and place them into the score exactly as you want it. (For example, it might interpret a held quarter-note as either a dotted quarter-note or one that's tied to a thirty-second note). Even though these computer algorithms are getting better at interpreting musical data, and quantization can be used to tell a computer to *round* a note value to a specified length, a score will still often need to be edited manually to correct for misinterpretations.

Before continuing, I'd like to point out the importance of taking the time to properly set up the initial default settings of a music notation program.

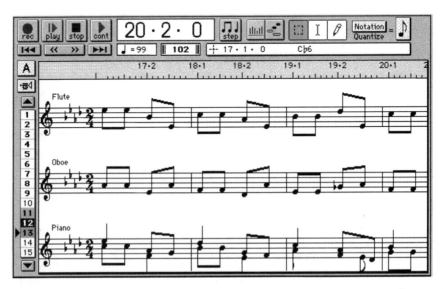

Figure 7.3   Opcode's Musicshop notation editing window. (Courtesy of Opcode Systems Inc., www.opcode.com)

These can be in the form of global settings that can help you to set such parameters as measure widths, number of stems, instrument layouts, and default key signature. These settings can even help take care of the less important decisions, such as title, composer, and instrument name fonts. As a time-saving feature, you might want to consider making a few general file templates that contain several of the more common time and key signatures, stave and instrument layouts, as well as other settings that you tend to work with.

Lastly, taking the time to familiarize yourself with the vast number of music notation parameters before you begin work on a song or score can save some real headaches later on when making changes (even minor ones that might impact a whole piece). Giving a composition some forethought can definitely help you avoid unnecessary bandage-type fixes at a later stage in the song.

Playing Back a Score

In the not-so-distant past, musicians commonly had to wait months and/or years to hear a finished composition. Orchestras and ensembles were generally expensive and often required the musician to have a financial patron or a good knowledge of corporate and/or state politics. One simple solution to this was the piano reduction, which served to condense a score down into a compromised version that could be played at the piano keyboard.

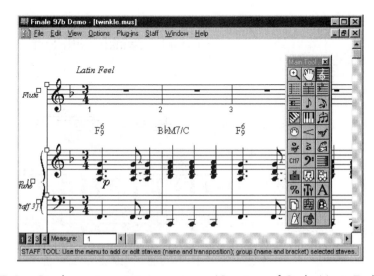

Figure 7.4   Finale music notation program. (Courtesy of Coda Music Technology, www.codamusic.com)

With the advent of MIDI, however, a score can now be played through various MIDI instruments in a composing/project studio, directly from the notation program (Figure 7.4). In this way, a system can approximate a working rendition of the final performance with relative ease. The ability to play back a composition also lets the artist check for final errors before being subjected to the expense and inherent difficulties of working with a live orchestra.

Printing a Score

Once the score has been edited into a final form, the process of creating hard copy to print is simple. Generally, a notation program will let you lay out the score in a way that best suits your taste, the producer's taste, or the score's intended purpose. Often a professional program will let you make final changes to such parameters as margins, measure widths, title, copyright, and other text-based information. Once completed, the final score can be printed on either a high-quality ink-jet or laser printer (Figure 7.5).

Printing from Your Sequencer

As we saw in Chapter 5, many newer sequencing packages will let you view and edit a single MIDI track using standard music notation (Figure 7.6). Although this option can be extremely useful for making edits using

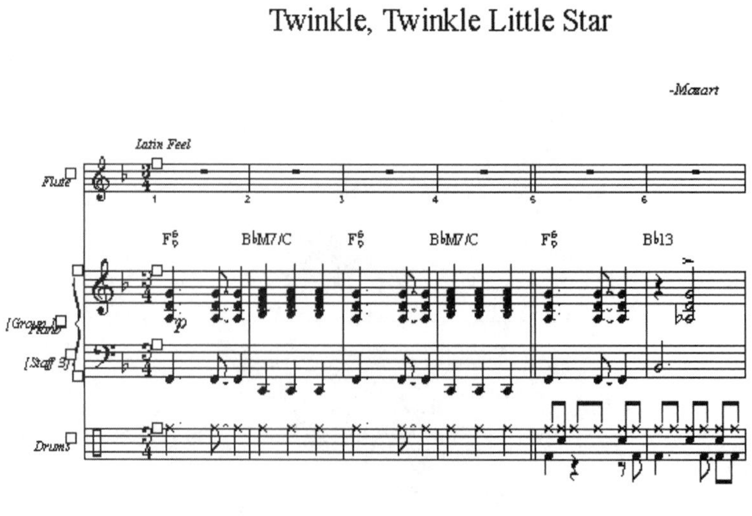

Figure 7.5 A printed score. (Courtesy of Coda Music Technology, www.codamusic.com)

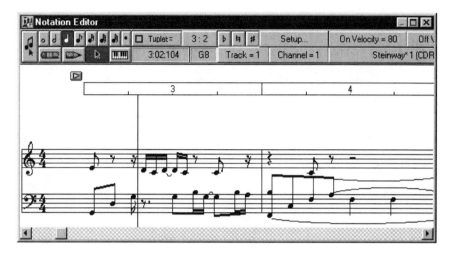

Figure 7.6 A sequencer notation window. (Courtesy of Passport Designs Inc., www.passportdesigns.com)

traditional music notation, a sequencer's ability to print a score or a track is rather limited when compared to the overall layout features of a music printing program. That's not to say that most sequencing packages won't do a good job of printing a simple part, because they can be great. I've had more than one session where the string parts have been printed directly from my sequencing program. Clearly, for those that don't have or don't need the flexibility of a professional music printing program, your sequencer's music printing option can often provide you with basic printouts and lead sheets that are just fine.

8

DIGITAL AUDIO IN MIDI PRODUCTION

Over the years, digital audio technology has grown to play a strong role in MIDI production. This is largely due to the fact that MIDI is a digital medium and as such can easily be interfaced with devices that output or control digital audio. Devices such as samplers, sample editors, hard disk recorders, and digital audio tape recorders (of the DAT, MINI Disc, and modular digital multitrack varieties) are commonly used to record, reproduce, and transfer sound within such an environment.

In recent years, the way that electronic musicians store, manipulate, and transmit digital audio has changed dramatically. As with most other media, these changes have been brought about by the integration of the personal computer into the modern-day project studio environment. In addition to sequencing MIDI data and controlling instruments in a MIDI system, newer generations of computers and their hardware peripherals have been integrated into the MIDI environment to receive, edit, manipulate, and reproduce digital audio with astonishing ease. This chapter is dedicated to the various digital system types and to the details of how they relate to the modern-day project studio.

Samplers

As we saw in Chapter 4, a *sampler* is a device that's capable of recording, musically transposing, processing, and reproducing segments of digitized audio directly from RAM memory. Because this memory is often limited in size (relative to digital audio's memory-intensive nature), the segments are generally limited in length and range from only a few seconds to one or more minutes.

Assuming that sufficient memory is available, you can load any number of samples into the system and then reproduce them from a controller. These samples can be transposed in real time, either up or down, over a number of octave ranges in a polyphonic fashion. Quite simply, this musical transposition occurs by reproducing the recorded sample files at sample rates that correspond to established musical intervals. These samples can then be assigned to a specific note on a MIDI controller or mapped across a keyboard range in a multiple-voice fashion. Extensive editing capabilities are available for modifying sounds in much the same way as a synthesizer can modify a wavetable sample.

In this day and age, an enormous number of previously recorded and edited sample files containing instruments, effects, noises, and so on, can be purchased on disk, CD, or CD-ROM. Such commercially available sample libraries are often the mainstay of both electronic musicians and visual post-production facilities. However, commercial sample libraries needn't be your only option. A number of professional and nonprofessional artists prefer to create their own sample libraries. These can be created from acoustically recorded or electronically generated sound sources, although I'd be remiss if I didn't say that samples are often "lifted" from previously recorded source material (such as CD, TV, records, and videotapes).

Sample Editing

When a recorded sound is transferred into a sampler, the original source material often contains extraneous sounds, such as breathing noises, fidget sounds, or other music that might occur both before and after the desired sound (Figure 8.1A). At this point, the unwanted sounds can be edited out by trimming their respective in-points and out-points, so that the system's microprocessor ignores all the samples before a user-defined in-point and/or following a desired out-point (Figure 8.1B). After trimming, the final sample can be played, processed, and saved to floppy or hard disk for later recall.

Looping

Another editing technique that's regularly used by sample artists to maximize the system's available RAM- and disk-based memory is a process known as *looping*. With the looping technique, a sample that occupies a

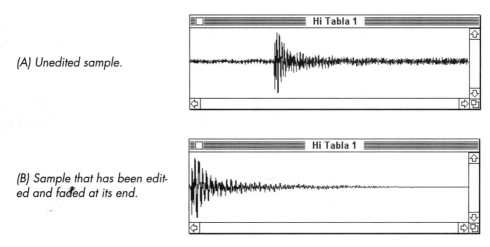

(A) Unedited sample.

(B) Sample that has been edited and faded at its end.

Figure 8.1   Basic sample editing.

finite memory space in RAM can be sustained for long periods of time, thereby preventing the sound from abruptly ending while the key on a MIDI keyboard is still held down.

Such a loop is created by defining a region in an audio segment that doesn't significantly change in amplitude and character over time and then repeatedly playing this section back from within RAM (Figure 8.2). Depending on the sound source, a loop can be created from waveform segments that are very short or, conversely, those that last several seconds.

When creating a looped splice, you can often make life easier by following this simple rule: *Match the waveform shape and amplitude at the beginning of the loop with the waveform shape and amplitude at its end.* This simply means that the waveform amplitude at the beginning of the loop must match the amplitude of the ending loop point (Figure 8.3). If these aren't matched, the signal levels will vary and an annoying "pop" or audible "tick" will result. Many samplers and sample editing programs provide a way to automatically search for the closest level match or display the loop crossover points on a screen, so that you can manually match amplitude levels.

Certain samplers will let you program more than one loop into a sample file. This has the effect of making the sample sound less repetitive and more natural and also adds to its overall expressiveness when played on a keyboard. In addition to having multiple sustain loops, a release loop can be programmed to decay the sample when the keyboard note is released.

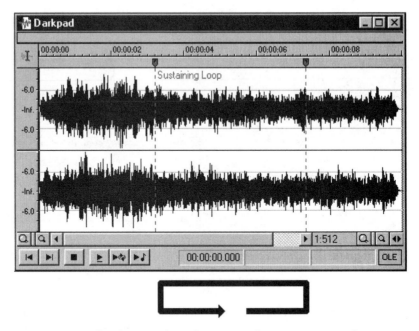

Figure 8.2   Example of a sample with a sustain loop. (Courtesy of Sonic Foundry, www.sonicfoundry.com)

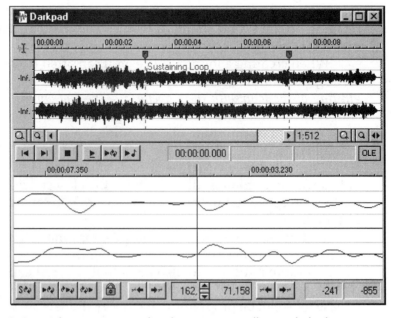

Figure 8.3   A loop tuning window lets you manually match the beginning and end levels of a loop. (Courtesy of Sonic Foundry, www.sonicfoundry.com)

The Sample Editor

During the past few years, several MIDI sample dump formats have been developed that let you transmit and receive samples in the digital domain. These sample dumps can be made between samplers or between a personal computer and a sampler. To take full advantage of the latter process, sample editing software (Figure 8.4) was developed to perform important sample-related tasks such as the following:

- Load samples into a computer, where they can be stored to hard disk, arranged in a library that best suits the user's needs, and transmitted to any sampling device in the system (often to samplers with different sample rates and bit resolutions).

- Edit samples and arrange the sample before copying to disk using standard computer cut-and-paste editing tools. Because most samplers support multiple loops, sample segments can be repeated to save valuable RAM memory.

- Process a signal to alter or mix a sample file digitally, using functions such as gain changing, mixing, equalization, inversion, reversal, muting, fading, crossfading, and time compression.

Distribution of Sampled Audio

Within a sample-based MIDI setup, sampled audio must be distributed in a way that's as fast and as painless as possible. Therefore, standards have been adopted that allow samplers and related management programs to communicate and store sample-based digital audio. Note that not all samplers make use of these protocols. As always, consulting the manual for

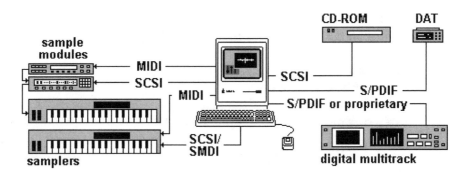

Figure 8.4 Example diagram of a sample editor network.

each device before attempting a data transfer will most likely reduce stress, frustration, and wasted time.

MIDI Sample Dump Standard

The *sample dump standard (SDS)* was developed and proposed to the MIDI Manufacturers Association as a protocol for transmitting sampled digital audio and sustain loop information between sampling devices as a series of MIDI SysEx messages. Although most samplers can communicate sample data via SDS, it has the distinct disadvantage of being rather slow, because the digital audio data is being transmitted over standard MIDI lines at the 31.25-kbaud rate. When transmitting anything more than a short sample, be prepared to take a coffee break.

SCSI Sample Dump Formats

A number of computer-based digital audio systems and professional samplers can also transmit and receive sampled audio via *SCSI (small computer systems interface)*. SCSI is a bidirectional communications line that's commonly used by personal computers to exchange digital data at high speeds. When used in digital audio applications, it provides a direct parallel data link for transferring sound files at a rate of 16 Mbytes/second (literally hundreds of times faster than MIDI) or higher. Although the data format will change from one device to the next (meaning that data can only be transmitted between like devices or a device and a PC that's installed with the proper device driver), SCSI still wins out as a fast and straightforward way to transfer data.

SMDI

The SCSI Musical Data Interchange (SMDI) was developed as a standardized, non-device-specific format for transferring digitally sampled audio between SCSI-equipped samplers and computers at speeds up to 300 times faster than MIDI's transmission rate of 31.25 kbytes per second. Using this format, all you need to transfer digital audio directly from one supporting sampler to another is to connect the SMDI ports using a standard SCSI cable and following the steps for transferring the sample.

Although SMDI is loosely based on MIDI SDS, it has more advantages over its slower cousin than just speed. For example, SMDI can distribute stereo or multichannel sample files. Also, it's not limited to files that are less than two megawords in length, and can transmit associated file information, such as filename, pitch values, and sample number range. Sound patch and device-specific setup parameters can also be transmitted and received over SMDI lines through the use of standard System Exclusive (SysEx) messages.

Hard Disk Recording

Once developers began to design updated sample editors, they discovered that through additional processing hardware, digital audio could be recorded and edited directly on a computer's hard disk. Thus, the concept of the hard disk recorder was born. These systems (Figures 8.5–8.7) serve as computer-based hardware and software packages for the recording, manipulation, and reproduction of digital audio data that resides on hard disk. These systems are commonly (but not necessarily) designed around a standard personal computer and its associated hardware.

The advantages of having a digital audio editing system in an audio production environment are numerous. The following list describes some of them:

- **Ability to handle longer sound files:** Hard disk recording time is limited only by the size of the hard disk (1 minute of 44.1 kHz stereo audio occupies 10.5 Mbytes of hard disk memory).

- **Random-access editing:** As audio is recorded onto hard disk, any point in a program can be accessed at any time, regardless of the order in which it was recorded. Nondestructive editing allows audio segments (often called *regions*) to be edited and played back in any context or order without changing or affecting the originally recorded sound file in any way.

- **DSP:** Digital signal processing can be performed on a segment or entire sample file in either real or non-real time in a nondestructive fashion.

In addition to these advantages, computer-based digital audio devices serve to integrate many of the tasks that are related to both digital audio and MIDI production. These hard disk recording systems offer a new degree of power

Figure 8.5
Windows Sound Recorder application. (Courtesy of Microsoft Corp., www.microsoft.com)

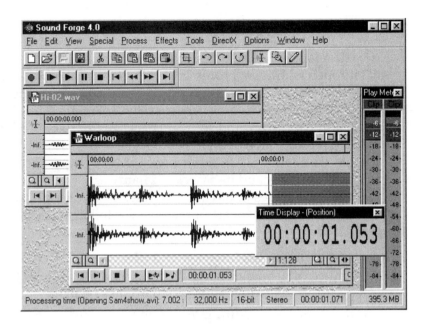

Figure 8.6 Sound Forge 4.0 digital audio editor (Courtesy of Sonic Foundry, www.sonicfoundry.com)

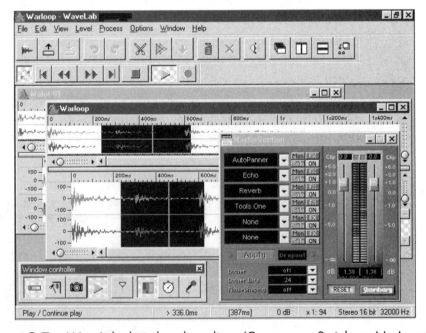

Figure 8.7 WaveLab digital audio editor. (Courtesy or Steinberg N. America, www.steinberg-na.com)

to the artist who relies on sound technology, because they can cost effectively offer unprecedented control, editing, processing, and connectivity to an existing production system.

Hard Disk Editing

In addition to recording and playing back long sound files, one of the strongest features of a hard disk recorder is its ability to perform extensive edits on a sound file compared to the time it would otherwise take to perform similar edits on an analog recorder. Furthermore, if you don't like the final results, you can easily "undo" a disk-based edit. You can even save an edited program, create a number of different versions, and then choose between them at a later time.

Nondestructive Editing

Nondestructive editing refers to a disk-based recorder's ability to edit a sound file without altering the data that was originally recorded to disk. This important capability means that any number of edits, alterations, or program versions can be performed and saved to disk without altering the original sound file data.

The nondestructive editing process is accomplished by accessing defined segments of a recorded sound file (often called *regions*) and outputting them in a user-defined order. In effect, when you define a specific region, you're telling the program to define a block of memory that begins at a specific memory address on the hard disk and continues until the ending address has been reached (Figure 8.8). Once defined, these regions can be inserted into the list (often called a *playlist*) in such a way that they can be accessed and reproduced in any order. Depending on the program, these playlists (also known as *editlists*) are built in the background and contain instructions about how to play back these regions. For example, Figure 8.9 shows a snippet from *Gone With the Wind* that contains the immortal words "Frankly, my dear, I don't give a damn." By segmenting it into three regions we could request that the editor output the words in several ways.

Destructive Editing

Destructive editing, on the other hand, occurs whenever the recorded data is altered and rewritten to disk in such a way that it can't be recovered in its original form. Obviously, this edit form is less desirable than its nondestructive counterpart, because edited data can't be easily recovered without saving each edited rendition.

Figure 8.8
Example of a defined
region.

I don't give a damn.

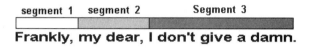

segment 1	segment 2	Segment 3

Frankly, my dear, I don't give a damn.

segment 2	segment 1	Segment 3

my dear, Frankly, I don't give a damn.

Figure 8.9
Example of how a playlist
could output Rhett's state-
ment from *Gone With the
Wind.*

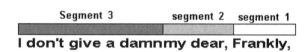

Segment 3	segment 2	segment 1

I don't give a damnmy dear, Frankly,

Although destructive editing isn't the preferred method, there are times
when it can be very useful. For example, you might want to save an edited
file as a single, separate sound file that can be easily retrieved and reproduced without additional playlist assembly or processing. For example, you
can save a complex series of mixed and edited files to disk as a single sound
effects file and then export it to a sampler, or simply play it from disk.

Basic Editing Techniques

One of the strongest assets of a hard disk recording system is its ability to
edit segments of digital audio with speed and ease. The following sections
offer a brief introduction to many of these valuable editing tools.

Graphic Editing

Most hard disk recording systems graphically display sound file information on a computer or LCD screen as a series of vertical lines that represent
the overall amplitude of a waveform over time in a WYSIWYG (what-you-
see-is-what-you-get) fashion (Figure 8.10). Depending on the system type,
sound file length, and the degree of zoom, the entire waveform can be shown
on the screen or only a portion can be shown with the waveform continuing to scroll off one or both sides of the screen.

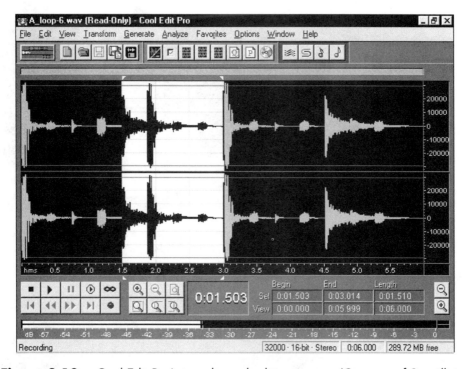

Figure 8.10 Cool Edit Pro's two-channel editing screen. (Courtesy of Syntrillium Software Corporation, www.syntrillium.com)

Graphic editing differs greatly from the "razor blade" approach used to cut analog tape in that it offers both visual and audible cues as to where a precise edit point will be. Using graphic editing, waveforms that have been cut, pasted, reversed, and assembled are visually reflected on the screen. Usually these edits are nondestructive, allowing the original sound file to remain intact.

Only when a waveform is "zoomed in" fully is it possible to see the individual waveshapes of a sound file (Figure 8.11). Often when a sound file display is zoomed in at this level, the individual samples can be redrawn to remove potential offenders (such as clicks and pops) or smooth amplitude transitions between loops or adjacent regions.

When working in a graphic editing environment, a defined sound file segment is commonly referred to as a *region*. In general, you can define a region by positioning the cursor within the waveform, pressing and holding the mouse or trackball button, and then dragging the cursor to the left or right. Usually, the selected region is then highlighted for easy identification (Figure 8.12). After the region is defined, it can be edited, marked, named, or otherwise processed.

135

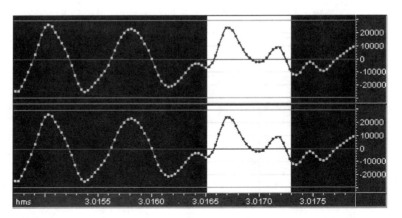

Figure 8.11 Often, when a sound file is fully zoomed in will the actual shape of a waveform be displayed. (Courtesy of Syntrillium Software Corporation, www.syntrillium.com)

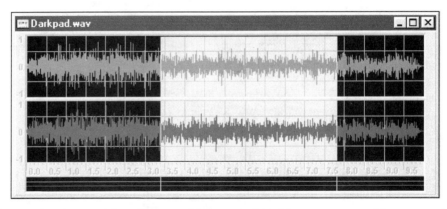

Figure 8.12 A defined region. (Courtesy of Goldwave, www.goldwave.com)

Cut-and-Paste Editing Techniques

Once you've introduced yourself to the world of graphic editing, the next step is to begin to cut and paste a sound file into a basic sequence that best fits the project. The basic cut-and-paste techniques used in hard disk recording are entirely analogous to those used on a word processor or any graphics-based program:

- **Cut:** Places the highlighted region into memory and deletes the selected data (Figure 8.13).
- **Copy:** Places the highlighted region into memory and doesn't alter the selected waveform in any way (Figure 8.14).

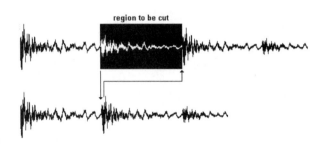

Figure 8.13
Cutting a waveform region.

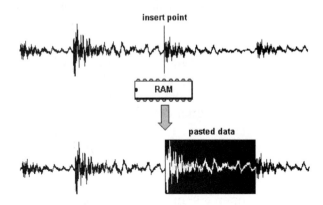

Figure 8.14
Copying a waveform region.

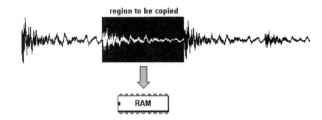

Figure 8.15
Pasting a waveform region from the Clipboard.

- **Paste:** Copies the waveform data that's within the system's clipboard memory into the sound file beginning at the current cursor position (Figure 8.15).

Digital Signal Processing

In addition to being able to cut, copy, and paste various regions of a sound file, you can also alter a sound file or segment using *digital signal processing (DSP)* techniques. In short, DSP works by directly altering the samples in a defined region according to a program algorithm (a set of programmed

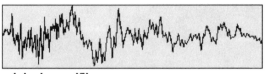

original soundfile

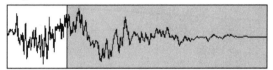

**defined region is faded and the
resulting data is written to disk
as a separate file**

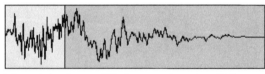

Figure 8.16
Performing a non-real-time
fade and writing it to disk
as a separate file.

**The original soundfile is reproduced until the
fade is reached, at which point the tagged file
containing the fade's data will begin to play.**

instructions), so as to achieve a desired result. This processing is often per-formed in an off-line fashion. This means that the processor most likely won't be available for playback or other processing tasks until the neces-sary signal calculations have been performed, and the resulting DSP cal-culations have been written to disk as a separate file that can be tagged as an original sound file. In this way, should the effect not be what you wanted, you could simply "undo" the effect (thereby removing the processed file from the edit playlist) and then reprocess the segment. As an example, let's open a short sample, fade out its end and then save the results as a sepa-rate file (Figure 8.16).

Do It Yourself Tutorial: Fading and Saving a Sample

1. Open or create a short sound file (3 to 4 seconds would do nicely).
2. Highlight a region that ranges from 1 second into the file to its end.
3. Instruct the system to perform a fade-out (a simple set of calculations that reduces the amplitude over time). Note that during playback, the hard disk recorder will play the original file up to the point where the fade was created and will seamlessly switch to the newly calculated fade without an audible break.
4. Save the file under a new name. This instructs the editor to write a whole new file that contains the fade and initial intro.

Real-Time Digital Signal Processing

Real-time signal processing differs from its non-real-time counterpart in that the algorithmic calculations are immediately carried out during playback without having to wait. Real-time DSP may require the use of extra processors in addition to the computer's CPU, although some programs are able to use the extra horsepower of newer processors to carry out real-time DSP.

Basic Digital Signal Processing Techniques

The following section details many of the basic DSP functions that can be found on a hard disk editing system.

Amplitude

Besides basic cut-and-paste techniques, processing the amplitude of a signal is one of the most common types of changes that you're likely to encounter. These include such processes as gain changing, fading, and normalization.

Gain changing relates to the altering of a region or track's overall amplitude level, whereby a signal can be reduced or increased proportionally to a specified level (often in decibels or percentage value). To increase a sound file or region's overall level or signal resolution, a function known as *normalization* can be used. Normalization (Figure 8.17) refers to changes in gain such that the overall signal is proportionately raised to a level whereby the greatest signal amplitude will be at 100% (full digital signal level). This function is often used to take full advantage of the audio hardware's dynamic range.

Fades and Crossfades

The fading in or fading out of a region is accomplished by increasing or reducing a signal's relative amplitude over a defined duration. For example, fading in a file (Figure 8.18A) proportionately increases a region's gain from infinity (zero) to full gain. Likewise, a fade-out has the opposite effect of

Figure 8.17
Original signal and normalized (full-gain) signal level.

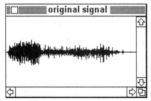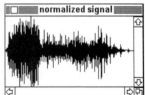

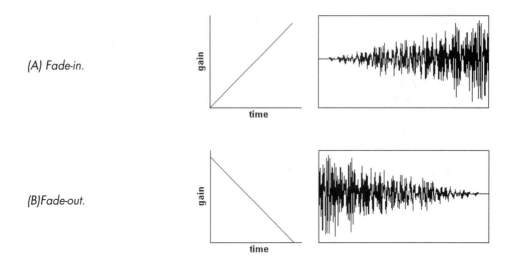

(A) Fade-in.

(B)Fade-out.

Figure 8.18　Examples of various fade curves.

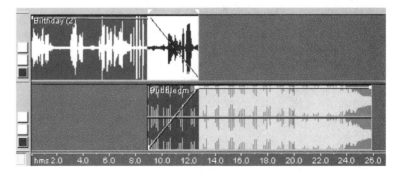

Figure 8.19　Cross-fades can be performed on a multichannel editor by overlapping files that are faded in and faded out. (Courtesy of Syntrillium Software Corporation, www.syntrillium.com)

creating a transition from full gain to infinity (Figure 8.18B). These DSP functions have the advantage of creating a much smoother transition than would otherwise be humanly possible during a manual fade.

A crossfade (or X-fade) is often used to smooth the transition between two audio segments that are either sonically dissimilar or don't match in amplitude at a particular edit point (a condition that often leads to an audible "click" or "pop"). Technically, the overall effect of this process is that the amplitude of the two signals are averaged over a user-definable length of time to mask the offending edit point (Figure 8.19).

Advanced Digital Signal Processing Editing Tools

Most hard disk recording systems offer editing functions beyond the basic cut-and-paste commands. However, the number of processing functions, the degree of complexity, and their flexibility often vary between systems. Depending on the system's capabilities, these functions can be performed in either real or non-real time.

Equalization

Digital EQ has become a common feature that varies in form, flexibility, and musicality from one system to the next (Figure 8.20). Most systems let you have full parametric control over the entire audible range and offer a variable bandwidth (Q) function. This feature lets you change the range of affected frequencies from being subtle and broadband, to tightly controlled and severe (i.e., a notch filter). An additional bonus that comes from using a software-based equalizer is that you can often alter a single region, single track, or the entire program with complete automation and repeatability.

Dynamic Range Processing

Dynamic range processors can be used to change the signal level of a program. Using algorithms that emulate a compressor (a device that reduces gain by a ratio that's proportionate to the input signal), limiter (reduces gain at a fixed ratio above a certain input threshold), or expander (increases the overall dynamic range of a program), the dynamics of a region, single track, or entire program can be adjusted (Figures 8.21 and 8.22).

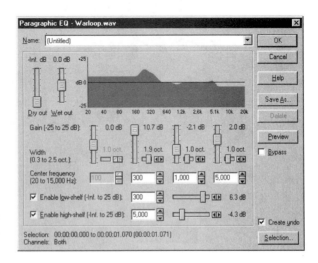

Figure 8.20
Paragraphic EQ screen. (Courtesy of Sonic Foundry, www.sonicfoundry.com)

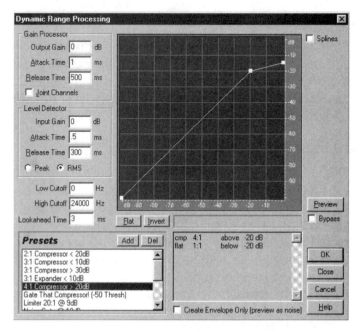

Figure 8.21 Cool Edit Pro's dynamic range processing screen. (Courtesy of Syntrillium Software Corporation, www.syntrillium.com)

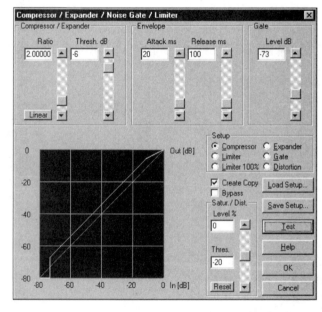

Figure 8.22 Samplitude's dynamic processing screen. (Courtesy of SEK'D, www.sekd.com)

(A) Digital audio can be shifted downward by interpolating, adding data to the original segment, and then reproducing the data at its original sample rate.

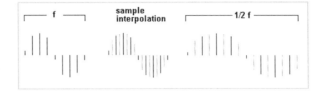

(B) Digital audio can be shifted upward by dropping samples from the segment and then reproducing this data at its original sample rate.

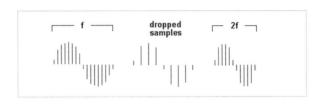

Figure 8.23   Basic disk-based pitch shifting techniques.

Pitch and Time Changing

The pitch change function lets you shift the relative pitch of a defined region or entire sound file either up or down by a specific percentage ratio or musical interval. Less expensive systems will often pitch shift by determining a ratio between the present and the desired pitch and then add (lower pitch) or drop (raise pitch) samples from the existing region or sound file (Figure 8.23).

In addition to being able to raise or lower the relative pitch, more sophisticated systems (Figure 8.24) can combine variable sample rate and pitch

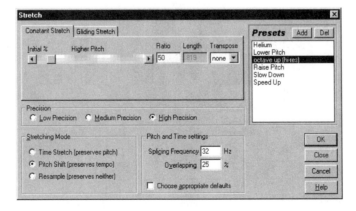

Figure 8.24   A time/stretch function window. (Courtesy of Syntrillium Software Corporation, www.syntrillium.com)

shift techniques to alter the duration of a region or sound file. Such system's can perform pitch- and time-related combinations such as:

- **Pitch shift only:** A program's pitch can be changed while its length remains the same.
- **Duration shift only:** A program's length can be changed while the pitch stays the same.
- **Pitch and duration shift:** A program's pitch can be changed while also having a corresponding change in length.

Digital Signal Processing Plug-Ins

One of the latest and most exciting additions to the concept of hard disk recording as a "studio-in-a box" is the *plug-in*. These software applications have come about as the result of the extreme popularity of certain hard disk recording platforms (most notably Digidesign's Pro Tools editor for the Mac and the DirectX PC media driver standard from Microsoft). As a result, third-party developers have begun to design and program specific applications that can perform almost any task under the sun. Usually (but not always), these tasks are beyond the manufacturing and development scope of the original software developer and, on many occasions, small businesses have

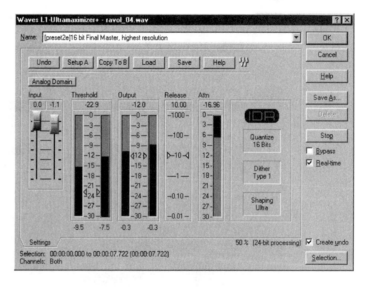

Figure 8.25   L1-Ultramaximizer peak-limiting plug-in is also capable of dithering (noise-shaping) and accurately requantizing between 20-, 16-, 12-, and 8-bit signals. (Courtesy of ks Waves, www.waves.com)

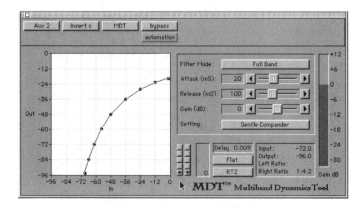

Figure 8.26 The MDT (Multiband Dynamics Tool). (Courtesy of Antares Systems, www.antares-systems.com)

sprouted up to create products based on market demand and good ideas. Examples of these plug-ins are detailed in Figures 8.25 through 8.29. Keep in mind, however, that the list of new developers is constantly growing—almost on a monthly basis.

Figure 8.27 Opcode fusion:VINYL adds vintage noises to a program. (Courtesy of Opcode Systems Inc., www.opcode.com)

145

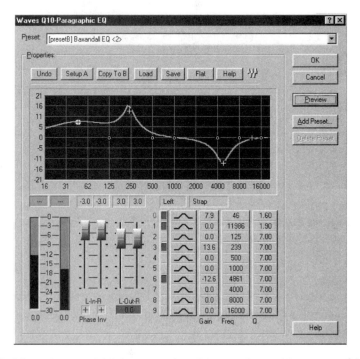

Figure 8.28 Waves Q10-Paragraphic EQ equalizer. (Courtesy of ks Waves, www.waves.com)

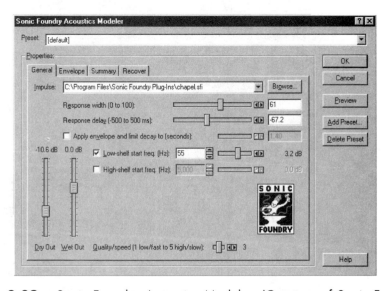

Figure 8.29 Sonic Foundry Acoustics Modeler. (Courtesy of Sonic Foundry, www.sonicfoundry.com)

Hard Disk Editing Systems

Two-channel and multichannel hard disk systems are in common use for sound effects, music editing, broadcast, Internet audio, and project studio recording. In fact, these hardware- and software-based systems can be found in almost every imaginable production environment because they're often fast and cost effective (Figures 8.30 and 8.31). Although a number of dedicated hard disk recording systems exist, the vast majority of systems use a standard multimedia or high-quality soundcard that can be plugged into an existing personal computer (Figures 8.32 and 8.33). Because most computers have integrated MIDI into their media environments, it often follows that MIDI sequencing can be integrated with digital audio to create a flexible and powerful production environment.

A large number of two-channel and multichannel systems are currently on the market and range from hardware card/software systems with prices that start at under $100 (such as the Soundblaster and similar consumer multimedia clones) to high-quality, sophisticated systems that are priced well into the thousands of dollars. When buying a system to integrate into your own production environment, you should stop and think about your personal needs. Is the quality sufficient? Will it integrate well into your system? What are the benefits/compromises and will it grow

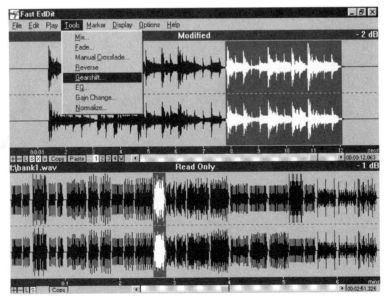

Figure 8.30 Fast EdDit hard disk editor for the PC. (Courtesy of Minnetonka Software, Inc., www.minnetonkasoftware.com)

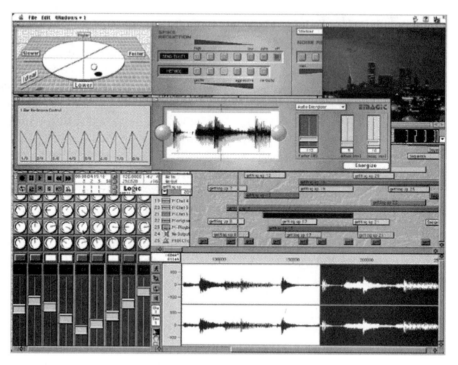

Figure 8.31 Logic Audio 3.0 integrated MIDI, music printing, and digital audio software. (Courtesy of Emagic, www.emagic.de)

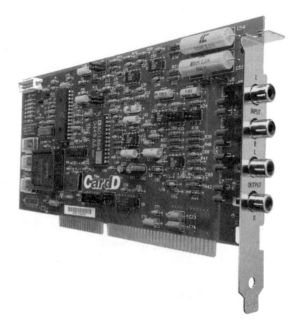

Figure 8.32
The CardD Plus soundcard.
(Courtesy of Digital Audio Labs,
www.digitalaudio.com)

Figure 8.33
The STUDI/O optical
ADAT/S/PDIF soundcard.
(Courtesy of Sonorus, Inc.,
www.sonorus.com)

with you as future system expansions occur? When researched thoroughly, a quality hard disk system can be a tremendously flexible, creative production tool that can grow as your system becomes more sophisticated.

The Virtual Track

As signal processing power, hard disk speeds, and overall hardware performance have increased, it has become easier for a hard disk system to
have additional processing overhead in reserve, when performing edit- and
mix-based tasks. This increased overhead has made it possible for numerous audio channels to be simultaneously accessed from a single two-channel
or multiple card system. This technological fact has brought about new technology in the form of the virtual multichannel hard disk editor.

In plain English, a virtual track system can access multiple audio channels (or stereo pairs) and then mix them, in real or non-real time, to a single
output channel or pair of channels (Figures 8.34 and 8.35). Such a system can
give you access to 8, 16, or more simultaneous channels on a Mac or PC-
based hard disk system, or it can mix multiple tracks down to your two-
channel soundcard.

At this point you might ask, "Why have all these channels if they are
just going to be mixed down?" Well, since the audio data can be defined
into regions that can be assigned to separate virtual "tracks," each sound can
be separately positioned into the project, processed, panned, mixed (in real
time), and muted in a number of ways without altering the original sound
file data on disk. Such nondestructive systems can be fully automated so

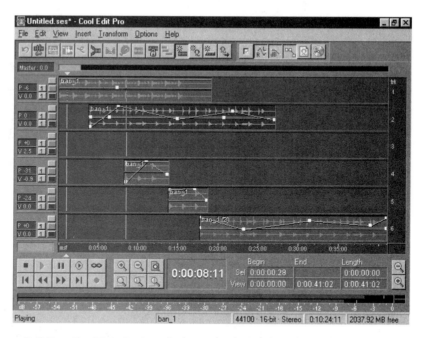

Figure 8.34   Cool Edit Pro's multichannel edit window. (Courtesy of Syntrillium Software Corporation, www.syntrillium.com)

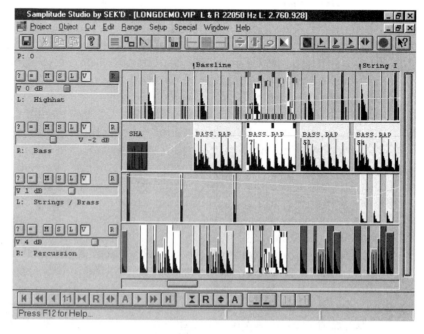

Figure 8.35   Samplitude's main editing screen. (Courtesy of SEK'D, www.sekd.com)

that all the necessary mix functions are saved to disk. In this way, newly as-sembled projects and mixes can be created any number of times until you've gotten the results you want.

Such a flexible, multichannel system is often useful in MIDI production because it makes it possible for you to integrally lock vocals, acoustic in-struments, effects, and other continuous sound files to a MIDI sequence. Each sound file or region can be edited, processed, looped, and slipped in time to easily match on-screen action or dialog to a visual media, or create station IDs, spots, and effects in a broadcast setting. The track and output structure of such a system can give you extensive flexibility over signal pro-cessing and mixing control in many types of audio production.

The Digital Audio Workstation

In recent years, the term *digital audio workstation (DAW)* has increasingly come to signify either a dedicated or computer-based hard disk recording system that offers advanced editing and signal processing features (Figures 8.36 and 8.37). Although a workstation can perform a wide range of audio-related functions, one of its greatest advantages is the distinct ability to in-tegrate a wide range of applications and devices into an audio production system. In effect, these systems are often able to integrate audio, video, and music media into a single, multifunctional environment that can freely com-municate data and perform tasks that relate to MIDI sequencing, sam-ple/playlist editing, sampling, hard disk recording, digital signal processing, synthesis/resynthesis, and music printing.

Figure 8.36
V8 hard disk editing work-station for the PC. (Cour-tesy of Digital Audio Labs, www.digitalaudio.com)

151

Figure 8.37
Pro Tools IV/hui hard disk
editing workstation for the
Mac. (Courtesy of Mackie
Designs, www.mackie.com)

Throughout music and audio production history, we've become used to the idea that certain devices were only meant to perform a single task: A recorder records and plays back, a limiter limits, and mixers mix. In response to this, I always liken digital audio workstations to a chameleon that can change its functional "colors" to match the task at hand. In effect, a digital audio workstation isn't so much a device as a systems concept that can perform a wide range of audio production tasks with ease and speed. Some of the characteristics that can (or should be) displayed by such systems include the following:

- **Integration:** One of the major functions of a dedicated or computer-based workstation is to provide centralized control over digital audio recording, editing, processing, and signal routing, as well as to provide both transport and/or time-based control over MIDI/electronic music systems, external tape machines, and videotape recorders.

- **Communication:** A workstation should be able to communicate and distribute digital audio data (such as AES/EBU, S/PDIF, SCSI, SMDI, and/or MIDI sample dump standards) throughout the system. Timing and synchronization signals (such as SMPTE time code, MTC, and MIDI sync) should also be supported.

- **Speed and flexibility:** Speed and flexibility are probably a workstation's greatest assets. After you become familiar with a particular system, most production tasks can be tackled in far less time than would be required using similar analog equipment. Many of the extensive signal processing features would simply be next to impossible to accomplish in the analog domain.

- **Expandability:** The "ideal" workstation should be able to integrate new hardware and/or software components into the system with little difficulty.
- **User-friendly operation:** An important element of a digital audio workstation is its central interface unit—you! The operation of a workstation should be relatively intuitive and should forego any attempt at obstructing the creative process by speaking "computerese."

When choosing a system for yourself or your facility, be sure to take all of these considerations into account. Each system has its own strengths and weaknesses. When in doubt, research the system as much as possible before you commit to it. Feel free to contact your local dealer for a test drive. Like a new car, purchasing a digital audio workstation can potentially be an expensive proposition that you'll probably have to live with for a while. Once you've made the right choice, you can get down to the business of making music.

Digital Recording Systems

In addition to hard disk recording systems, one of the most important digital audio devices that can be found in a project studio are stereo and multitrack recorders. These devices are used everyday in conjunction with MIDI and hard disk digital audio to give you quick access to physical tracks, more audio tracks with which to work, and a medium for mixing your project down to a final format.

Digital Audio Tape

Without doubt, the digital audio tape (DAT) recorder is the standard medium for transferring partial or final mixes to tape with digital CD quality in both the recording and project studios. This is due to its high quality, reliability, and overall cost effectiveness. It is also important in MIDI production because it's a cost-effective device for recording and archiving your own samples. Because DATs work in the digital domain, sound files can be easily transferred to and from a sample or hard disk editor.

DAT technology (Figure 8.38) uses an enclosed compact cassette that's even smaller than a compact audio cassette. Equipped with both analog and digital input/outputs, the DAT format can record and play back at three standard sampling frequencies: 32, 44.1, and 48 kHz (although sample rate

153

Figure 8.38
Tascam DA-P1 portable
DAT recorder. (Courtesy of
TEAC Corporation,
www.tascam.com)

capabilities and system features may vary from one DAT recorder to the next). When using a consumer DAT machine, audio often can't be recorded at the 44.1-kHz sampling frequency using the digital inputs (and sometimes the analog-ins as well). This digital "block-out," known as the SCMS copy code system, was originally designed to discourage the unlawful copying of prerecorded DAT tapes and CDs. Fortunately, most professional machines don't use SCMS and can record at all sample rates.

Current DAT tapes offer running times of up to 2 hours when sampling at 44.1 and 48 kHz and reserve three record/reproduce modes at the 38-kHz sampling rate. Option 1 provides 2 hours of maximum recording time with 16-bit linear quantization. Option 2 provides up to 4 hours of recording time with 12-bit nonlinear quantization. Option 3 (which is never used) allows the recording of four-channel, nonlinear 12-bit audio.

DAT Tape/Transport Format

The actual track width of a DAT format's helical scan can range downward to about 1/10th the thickness of a human hair, allowing a density of 114 million bits per square inch. This is the first time such a density has ever been achieved using magnetic media. To assist with tape tracking, a sophisticated correction system is used to center the heads directly over the tape's scan path.

The head assembly of a DAT recorder uses a 90° half-wrap, helical tape path (Figure 8.39A). The term *helical* refers to the fact that the tape path is encoded at a slant across the thin tape path. This also means that the signal is recorded onto tape in digital bursts as one slant track is recorded, then the next, and so on (Figure 8.39B), necessitating the use of a digital buffer. On playback, these bursts are again smoothed into a continuous data stream. This encoding method has the following advantages:

- Only a short length of tape is in contact with the drum at any one time. This reduces tape damage and allows a high-speed search to be performed while the tape is in contact with the head.
- Low tape tension ensures a longer head life.

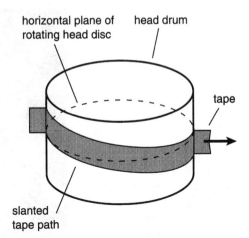

(A) *Half-wrap path, showing 90° tape contact.*

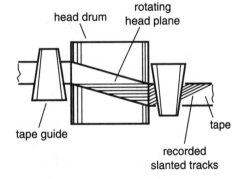

(B) *Basic detail of a helical scan.*

Figure 8.39
Helical scan tape path.

A high-speed search function (approaching 300 times normal speed) is a key feature of the DAT format. In addition to this, subcode information can be written into the digital bit stream, which serves as a digital identifier, just as a compact disk uses indexes for selection and timing information. These subcodes can be written as any one of three data types: start ID (indicates the beginning of a selection), skip ID (indicates a selection should be skipped over), and program number. In addition to these identifiers, subcode data can be used to encode time-related information and even SMPTE time code onto tape.

Modular Digital Multitrack Systems

One of the most monumental developments in recent recording history has been the introduction of the modular digital multitrack system (MDM for short). MDMs (Figure 8.40) are cost-effective multitrack digital audio

155

Figure 8.40
Studio equipped with an
MDM system. (Courtesy of
Alesis Studio Electronics,
www.alesis.com)

recorders that can record eight tracks of digital audio onto standard videotape cassettes.

These recorders are modular because they can be linked together, allowing additional tracks to be added to the system (in blocks of 8), with a
theoretical maximum limit of up to 128 tracks! If the capability to create and
expand a digital audio recording system to suit your needs isn't enough,
you might want to check out the price of an MDM. As of this writing, a basic
8-track digital MDM can be purchased for less than $2,000 (often signifi-
cantly less). Before these devices, the only available digital option would
cost more than $20,000 (often significantly more). So it's easy to see why
these modular, expandable systems have begun to revolutionize the music
industry—both in the home and in the studio. Currently, there are two popular MDM formats: the ADAT and DTRS (8-mm) formats.

ADAT

The ADAT format (established by the Alesis Corporation) is comprised of
an 8-track digital recorder that uses standard S-VHS videotape and features
a 16- or 20-bit resolution at sample rates of 44.1 and 48 kHz. When using a
standard 120-minute S-VHS tape at the highest sampling rate, an ADAT
tape's total recording time can range up to 1 hour.

Depending on the model and manufacturer, the analog input/output
(I/O) connections can be 1/4-inch phone, phono, or even XLR; however, a
standard 56-pin Elco connector is always available to connect all I/Os to a
mixer or other device, using a single multicable "snake."

Digital I/O connections use a proprietary fiber optic cable link that can
carry all 8 channels over a single "lightpipe." This link is used to connect an

Figure 8.41
Alesis BRC unit for the
ADAT. (Courtesy of Alesis
Studio Electronics,
www.alesis.com)

ADAT to such digital peripheral I/O devices as an S/PDIF or AES/EBU digital interface, professional audio interface card, or hard disk recording system.

Alesis remote functions are carried out through the use of the LRC (Little Remote Control) or a BRC (Big Remote Control). The LRC is shipped with every Alesis ADAT and includes all the device's front panel transport functions on a single, palm-sized remote. The optional BRC (Figure 8.41) can act as a full-featured remote control, digital editor, expanded-feature autolocator, and synchronizer on a single table-top or free-standing surface. It can control up to 16 remote ADAT units and bounce data from one track to another in the digital domain (while also allowing tracks to be shifted in location).

DTRS

The digital tape recording system (DTRS) format (established by the Tascam Corporation) is an 8-track MDM that can record up to 108 minutes of digital audio onto a standard 120-minute Hi-8mm videotape. Similar to the ADAT format, the DTRS system (Figure 8.42) can be expanded (in 8-track increments) up to a maximum of 128 tracks.

Figure 8.42
Tascam DA-98 digital multitrack recorder. (Courtesy of
TEAC Corporation,
www.tascam.com)

Depending on the manufacturer and model, analog I/O is commonly made via –10-dBV unbalanced RCA phono jacks, as well as balanced +4-dB connections (using two 25-pin D-sub multipin connectors). Digital I/O is made through Tascam's proprietary TDIF-1 (Tascam Digital Interface), which uses a 25-pin D-sub connector to link to external interface accessories or to another MDM for making digital clones.

External synchronization to time code can be designed into the machine or handled from an optional interface card that offers SMPTE chase-lock sync, MIDI machine control, Sony 9-pin RS-422 video editor control, and video I/O sync on standard BNC jacks.

Digital Transmission

In this digital age, it has become commonplace for digital audio data to be distributed from one device to another or throughout a connected production system in the digital domain. Using this medium, digital audio information can be transmitted in its original numeric form and thus, in theory, won't suffer from degradation whenever copies are made from one generation to the next.

When looking at the differences between the distribution of digital audio and analog audio, you should keep in mind that, unlike its counterpart, the transmitted bandwidth of digital audio data occurs in the megahertz range. Therefore, digital audio transmission has more in common with video signals than the lower bandwidth analog audio range. This means that impedance must be much more closely matched and that a quick-fix solution (such as using a "Y-cord" to split a digital signal between two recorders) is a major "No-No." Failure to follow these precautions could seriously deform the digital waveform.

Because of these tight signal restrictions, several digital transmission standards have been adopted that uniformly allow for the quick and reliable transmission of digital audio between devices that support these standards. The two most commonly encountered formats are the AES/EBU and S/PDIF protocols.

The AES/EBU (Audio Engineering Society and the European Broadcast Union) protocol has been adopted for the purpose of transmitting digital audio between professional digital audio devices. This standard is used to convey two channels of interleaved digital audio through a single, three-pin XLR microphone cable. This balanced configuration connects pin 1 to the signal ground, while pins 2 and 3 are used to carry signal data. AES/EBU transmission data is low impedance in nature (typically 110 ohms) with waveform amplitudes that range between 3 and 10 volts. These factors allow for maximum cable lengths of up to 328 feet (100 meters) without encountering undue signal degradation.

Digital audio channel data and subcode information is transmitted in blocks of 192 bits that are organized into twenty-four 8-bit words. Within the confines of these data blocks, two subframes are transmitted during each sample period that convey information and digital synchronization codes for both channels in a L–R–L–R . . . fashion. Because the data is transmitted as a self-clocking biphase code, wire polarity can be ignored, and the receiving device will receive its reference clock timing from the digital source device.

S/PDIF

The S/PDIF (Sony/Phillips Digital Interface) has been adopted for transmitting digital audio between consumer digital audio devices; it is similar in data structure to its professional counterpart. Instead of using a balanced three-pin XLR cable, the S/PDIF standard has adopted the single conductor, unbalanced phono (RCA) connector, which conducts nominal peak-to-peak voltage levels of 0.5 volts with an impedance of 75 ohms. In addition, transmissions via optical lines that use the Toslink optical connection cable generally use the S/PDIF data protocol.

As with the AES/EBU protocol, S/PDIF channel data and subcode information is transmitted in blocks of 192 bits; however, they are organized into twelve 16-bit words. A portion of this information is reserved as a category code that provides the necessary setup information (sample rate, copy protection status, and so on) to the copy device. A portion of the 24 bits set aside for transmitting audio data is used to relay track indexing information, such as start ID and program ID numbers, allowing index information to be transferred from the master to the copy. Note that the professional AES/EBU protocol can't digitally transfer these codes during transfer.

SCMS

Initially, the DAT medium was intended to provide consumers with a way of making high-quality digital recordings for their own personal use. Soon after its inception, however, for better or for worse, the recording industry began to see this new medium as a potential source of lost royalties due to home copying and piracy practices. As a result, the RIAA (Recording Industry Association of America) and the former CBS technology center set out to create a "copy inhibitor." After certain failures and long industry deliberations, the result of these efforts was a process that has come to be known as the Serial Copy Management System or SCMS.

SCMS (pronounced "scums") has been implemented in many consumer digital devices to prohibit the unauthorized copying of digital audio at the 44.1-kHz sample rate (standard CD rate). It doesn't apply to the making of analog copies, to digital copies made using the AES/EBU protocol, or to sample rates other than 44.1 kHz.

So what is SCMS? Technically, it's a digital protection flag that is encoded in byte 0 (bits 6 and 7) of the S/PDIF's subcode area. This flag can have only one of three possible states:

- **Status 00:** No copy protection, allowing unlimited copying and subsequent dubbing.
- **Status 10:** No more digital copies allowed.
- **Status 11:** A single copy can be made of this product, but that copy can't be copied.

Suppose that we have a CD player with an optical output and two consumer DAT machines equipped with SCMS. If we try to digitally copy a CD that has a 10 SCMS status, we'd simply be out of luck. But suppose we found a CD that has an 11 status flag? By definition, the bit stream data would inform the initial DAT copy machine that it's OK to record the digital signal. However, the status flag on the copy tape copy will then be changed to a 10 flag. If at a later time we were to clone this DAT copy, the machine doing the second-generation copy wouldn't allow itself to be placed into record. At this point, we have two possible choices: Record the signal using the analog ports (often with a minimum of signal degradation) or purchase a digital format converter that (among other things) lets us strip the SCMS copy protection flags from the bit stream and continue making multigeneration copies.

Signal Distribution

Both the AES/EBU and S/PDIF digital audio signals can be distributed from one digital audio device to another in a daisy chain fashion (Figure 8.43). This works well if only a few devices are chained together. However, if a number of devices are connected, time-base errors (known as jitter) may be introduced into the path, with the possible side effects being added noise and a "blurred" signal image. One way to reduce potential time-based errors is to use a digital distribution device that can route data from a single digital source to a number of individual device destinations (Figure 8.44).

Figure 8.43
Digital audio can be distributed in a daisy chain fashion.

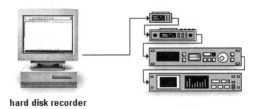

hard disk recorder

Figure 8.44
A distribution system can be used to route digital audio data to individual devices.

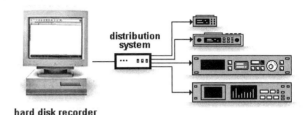

distribution system

hard disk recorder

CD Recorders

I simply couldn't let this chapter end without mentioning a device that has changed the way of life for both electronic musicians and tech-heads, alike. I'm referring, of course, to the CD-recorders (CD-Rs) and CD-writable recorders (CD-RWs) that have taken the industry by storm, by allowing us to record (a process know as "burning") our own audio CDs, as well as to back up our precious program and media files onto reliable and removable media.

Although most CD-Rs come bundled with software for burning both data and "Red Book" CD audio to disk, many of these programs are fairly basic in their operation and don't offer many advanced options. This is often fine when it comes to burning file-related data to disk; however, when burning audio files onto a disk, simple options (like pause times between tracks and real- or non-real-time level and fade adjustments) just are not available. When advanced options like these are required, you may have to purchase CD burning software from a third party (Figures 8.45 and 8.46).

When it comes to recording and backing up data to a CD, you should be aware of an important concept known as *packet writing*. As many of you who have CD-Rs know, until recently, everyone had to deal with the ISO-9660 standard of writing data to a disk all in one session (disk-at-once) or within several sessions (using multisession technology). This generally meant that you'd either have to save up enough data in one giant

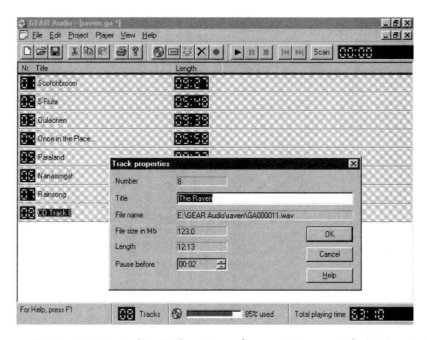

Figure 8.45　GEAR Audio CD burning software. (Courtesy of Electroson, Inc., www.Elektroson.com)

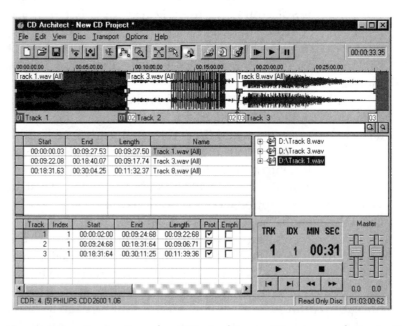

Figure 8.46　Sonic Foundry CD Architect. (Courtesy of Sonic Foundry, www.sonicfoundry.com)

directory to burn a CD in a single sitting or, worse yet, you'd have to deal with the inadequacies of burning multiple sessions.

With packet writing, all of the rules have changed. This CD burning technology allows small packets of variable or fixed-length data to be added to a CD-R disk "one file at a time." In addition to this, the disk's file table is updated each time data is written to disk, instead of being rewritten in its entirety. Let's take a look at what this techno-talk really means in the following paragraphs.

The most important part of packet writing is that it turns the CD-R into an actual drive in your system. "Drive letter" access means that the CD-R appears to the operating system as another drive letter, and any program that can write files to a drive can transparently write them to CD. Overwriting existing files is also the same as with other drives, with the obvious exception that you can't recover the previously used space when using a CD-R (although you can with a CD-RW).

Gone is the need to use a special program to write data to disk. Simply drag and drop the selected file or directory to your CD-R (just as you would with any other drive) or save the program file directly to disk. Because the data "packets" are smaller than the CD-R's memory buffers, it's impossible to have a buffer underrun error! This means that you can even copy files from a floppy directly to your CD without any problem. Up to now, this simply wasn't possible.

For more information on this amazing technology, contact Adaptec (www.adaptec.com) or OSTA (www.osta.org), an international trade association that's dedicated to promoting the use of writable optical technology for storing computer data and images. It's definitely worth checking out!

9

MULTIMEDIA

It's no secret that modern-day computers have gotten faster, sleeker, and sexier in their overall design. Their hardware and software systems have gotten more sophisticated (although some would add overly complex and crash-prone to this list). However, one of the crowning achievements of the modern computer is the degree of software and media integration that has come to be universally known by the household buzz word *multimedia*.

The Multimedia Environment

Basically, multimedia is nothing more than a unified programming and operating system (OS) environment that lets multiple forms of program data and playback media coexist and be routed directly to the appropriate hardware device for output, playback, and/or processing (Figure 9.1).

Although numerous complex design considerations must be in place if a multimedia computer is to work properly, once you understand two basic concepts, the rest will be much easier to understand. These are the concepts of task switching and the device driver.

Task Switching

Basically, task switching is a modern-day form of delusion. Just as a magic trick can be quickly pulled off with a sleight of hand or a film that switches frames 24 times each second can bring about the illusion of continuous movement, the multimedia environment deceives us into thinking that all

165

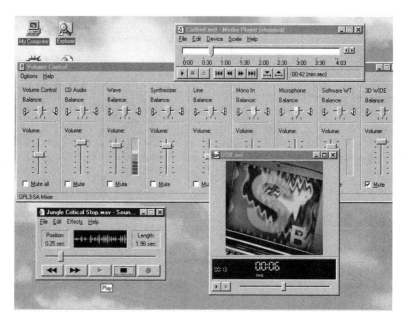

Figure 9.1   Example of a multimedia program environment.

of the separate programs and media types are working at the same time. In reality, the computer uses *task switching* to alternate its attention from one program to the next in a continuous fashion. Just like the film example, newer PCs have gotten so fast at cycling between programs and applications that we have the illusion that they are all running at the same time.

Device Driver

Another important concept is that of the *device driver*. In short, a "driver" acts as device-specific software patch cord that routes media data of a specific type to the appropriate hardware output device (also known as a *port*), and vice versa (Figure 9.2).

Figure 9.2
Basic interaction between a software application and hardware device via a device driver.

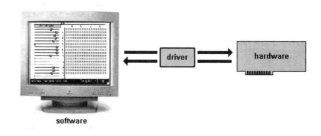

From all of this, a multimedia computer can be seen as being a device that operates and switches between multiple programs and media players and then, on receiving or playing back any media type, routes this data to/or from an appropriate hardware port. In short, these devices can simultaneously deal with and direct various forms of media at speeds that are so fast as to be virtually seamless. Cool, huh?

Hardware

Just as there are lots of computer types and levels of complexity, there is an equally wide variety of options for configuring the hardware of a computer to meet the demands of multimedia production. As is often the case, these hardware choices depend on the application. For the beginner or for general home applications, entry-level soundcards are often more than sufficient. These soundcards are commonly designed to handle simultaneously digital audio, MIDI, and FM or wavetable synthesis. Often, new computer systems are shipped with these cards already installed, or in the case of the Macintosh computer, most newer laptops, and some newer desktop PCs, these capabilities are built right into the system. The main drawback of most of these soundcard systems is their lack of quality from a music production standpoint, although they're often fine for basic media production. (In fact, most professional multimedia producers prefer Soundblaster-type cards, because most of their audience will be using this type of card.)

For those who will be producing music or high-quality media, systems can be assembled from a wide range of cards and hardware systems. These can range from a simple, high-quality audio card that costs slightly more than $100 to full-blown multichannel systems that can range well into the thousands of dollars.

A similar situation exists for MIDI. The basic PC soundcard design often includes a 16-channel (single-port) MIDI interface and a rather chintzy-sounding FM synthesizer. Medium-cost cards might offer two MIDI ports (32 channels) and include a higher quality wavetable synth; however, most professional systems that are intended for music production will have a dedicated soundcard and a multiport MIDI interface that can access eight or more ports, with literally hundreds of MIDI channels.

In the final analysis, the choice of hardware is up to you, as are the ways that you might want to apply them. It's always a safe bet to look at what your production needs are and then research the equipment that will fit those needs.

The Media

The basic media types grouped under the heading of multimedia are graphics, video, MIDI, digital audio, and the Internet (which uses all types of media). The following sections are intended to give you a basic overview of these types and how they are applied.

Graphics

Graphic imaging occurs on the computer screen in the form of pixels. These are basically little tiny dots that blend together to create color images in much the same way that dots are combined to give color and form to your favorite comic strip. The only differences are that the resolution is often greater on your color monitor, and the dot size doesn't change.

In a similar fashion to the way that word length affects the overall amplitude range of a digital audio signal, the number of digits in a graphic signal word will affect the total number of colors that each pixel can assume. For example, a 4-bit word has 16 possible combinations. Thus, a 4-bit word will allow your screen to have a total of 16 possible colors. Likewise, an 8-bit word will yield 256 colors, 16 bits will give you 65,536, and a 24-bit word length will yield a whopping total of 16.7 million colors!

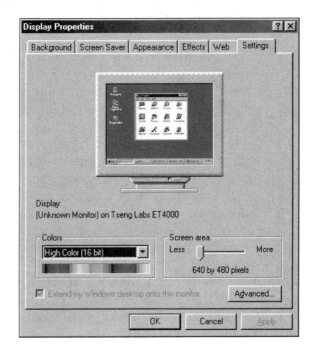

Figure 9.3
The Windows 98 Display Properties box is used to change screen sizing and pixel resolution.

The capabilities of your graphics display depend on your system's hardware (Figure 9.3), whereas the ability to create graphics depends on your hardware applications. Newer programming applications add a degree of "smarts" to your system's graphic capability, in that they can add shading, change lighting, alter color—all in real time according to a set of graphic parameters. This type of interactivity applies mostly to newer, CD-ROM games that can be played on either the Windows or Mac operating systems.

Desktop Video

Video lets us view images on our screen with full motion (Figure 9.4) and as such has become an integral part of multimedia. Basically, video is encoded onto the screen as a continuous series of frames that are refreshed at rates that vary from 12 or fewer frames/sec to the standard video rate of 30 frames/sec. You might think that these are extremely slow rates when compared to the much higher sampling rates of digital audio, but when you stop to think that each pixel in a video frame must be encoded into hundreds of thousands (if not millions) of pixel-related data during each frame, you'll begin to realize that the actual data required to encode and play back digital video is much higher. As a result, digital video is often compressed using a program algorithm that reduces the data rate by a significant amount. However, even with compression digital video is generally limited in either size or frame rate.

Figure 9.4
Example of a digital video window.

The two most common video standards are Quicktime for both the Mac and the PC and Microsoft's .AVI (audio/video interleaved) for the PC. Both the Microsoft and Macintosh OS platforms include a built-in application that allows these file types to be played without additional hardware or software.

MIDI

One of the unique advantages of MIDI as it applies to multimedia is the rich diversity of musical instruments and program styles that can be played back in real time, while requiring almost no processing overhead from the computer's CPU. This makes MIDI a perfect candidate for playing back soundtracks from multimedia games, Internet media, whatever. You could even imbed a soundtrack within a document (Figure 9.5).

It's interesting that MIDI has only now begun to take its rightful place as a serious music playback format for multimedia. Most likely, this delay in acceptance was due to a misunderstanding of the media and to the existence of a market that's still riddled with poorly designed FM synthesizers.

Fortunately, Microsoft has taken up the banner of imbedding MIDI within their media projects and has helped push MIDI onto the web. As a result, it's much more common for your PC to begin playing back a MIDI score, over (or perhaps in conjunction with) the more data intensive playback of soundtrack audio files.

Standard MIDI Files

The accepted format for transmitting files or real-time MIDI information in multimedia is the *standard MIDI file*. This file type (which is stored using the .MID or .SMF extension) is used to distribute MIDI data, song, track, time signature, and tempo information to the general masses. Standard MIDI files can support both single and multichannel sequence data, and can be loaded to, edited in, and then directly saved from almost every sequencer package.

When exporting a standard MIDI file, you may remember from an earlier chapter that they come in two basic flavors: type 0 and type 1. Type 0 is

Figure 9.5
Microsoft Windows Media Player application.

used whenever you want all of the tracks in a sequence to be compressed into a single MIDI track. All of the original channel messages will still reside within that track; it's just that the data would have no definitive track assignments. This data type might be the one to choose when creating a MIDI sequence for the Internet, where the sequencer or MIDI player application might not know how to deal with multiple tracks. The type 1 format, on the other hand, will most likely retain its original track structure and can be imported into another sequencer type with its basic track information and assignments intact.

General MIDI

One of the most interesting aspects of MIDI production is the absolute uniqueness of each professional and even semiprofessional project studio. In fact, no two studios will be alike (unless they've been specifically designed to be the same, or there's some amazing coincidence). Each artist will have his or her own favorite music equipment, supporting hardware, and favorite way of routing channels, tracks, and assigning patches. The fact that each system is unique has placed MIDI at odds with the need for absolute compatibility between systems in the world of multimedia. For example, upon importing a standard MIDI file over the net and loading it into a sequencer, you might hear a song that's being played with a set of totally irrelevant sound patches. (It might sound interesting, but it won't sound anything like the original artist intended.) Upon loading the MIDI file into a new computer, the sequence would again sound totally different, with patches that are so irrelevant that the guitar track might sound like a bunch of machine gun shots from the planet Glob.

To eliminate (or at best reduce), the basic differences that exist from system to system, a standard known as *General MIDI (GM)* was created. In short, General MIDI assigns a specific instrument patch to each of the 128 available program change numbers. Because all instruments that conform to the GM format use these patch assignments, placing these standardized program change commands at the header of each track will automatically configure the sequence to play with its originally intended instruments. Using this system, the differences from one multimedia synth to the next have finally been minimized. As an example, a standard MIDI file that conforms to General MIDI might contain the tracks shown in Table 9.1. Regardless of what sequencer is used to play back the file, as long as the receiving instrument conforms to the GM spec, the sequence will be heard using its intended instrumentation.

Table 9.2 details the program numbers and patch names that conform to the GM format. These patches include sounds that imitate synthesizers,

Table 9.1   Example of a General MIDI Sequence

Track	Prog. Change #	Instrument Name
1	33	Acoustic Bass
2	1	Acoustic Grand Piano
3	76	Pan Flute
4	47	Orchestral Harp
.		
10		Percussion
.		
16		

ethnic instruments, or sound effects that have been derived from early Roland synth patch maps. Although the GM spec states that a synth must respond to all 16 MIDI channels, the first 9 channels are reserved for instruments, while GM restricts percussion to MIDI channel 10 (Table 9.3).

Digital Audio

Digital audio is obviously a component that adds greatly to the multimedia experience. It can augment a presentation by adding a dramatic music soundtrack, help us to communicate through speech, or give realism to a soundtrack by adding sound effects.

Due to the large amount of data required to pass video, graphics, and audio from a CD-ROM, the Internet, or other media, the bit rate and sample rate structure of an audio file are usually limited compared with that of a professional quality sound file. The general accepted sound file standard for multimedia production is either 8-bit or 16-bit audio at a sample rate of 11.025 or 22.050 kHz. This standard has come about mostly because older single- and two-speed CD-ROMs generally couldn't pass the professional 44.1-kHz sample rate with full motion video or other graphics types without encountering annoying and spazmatic interruptions. In addition, larger, pro-rate sample files could take minutes or even hours to download over the Internet. As a result, the multimedia industry has pretty much decided to let the audio soundtrack take a backseat to other media forms. Fortunately, with improvements in compression techniques, hardware speed, and design, the overall sonic and production quality of sound has greatly improved.

Table 9.2   GM Instrument Patch Map

1. Acoustic Grand Piano	33. Acoustic Bass	65. Soprano Sax	97. FX 1 (rain)
2. Bright Acoustic Piano	34. Electric Bass (finger)	66. Alto Sax	98. FX 2 (soundtrack)
3. Electric Grand Piano	35. Electric Bass (pick)	67. Tenor Sax	99. FX 3 (crystal)
4. Honky-tonk Piano	36. Fretless Bass	68. Baritone Sax	100. FX 4 (atmosphere)
5. Electric Piano 1	37. Slap Bass 1	69. Oboe	101. FX 5 (brightness)
6. Electric Piano 2	38. Slap Bass 2	70. English Horn	102. FX 6 (goblins)
7. Harpsichord	39. Synth Bass 1	71. Bassoon	103. FX 7 (echoes)
8. Clavi	40. Synth Bass 2	72. Clarinet	104. FX 8 (sci-fi)
9. Celesta	41. Violin	73. Piccolo	105. Sitar
10. Glockenspiel	42. Viola	74. Flute	106. Banjo
11. Music Box	43. Cello	75. Recorder	107. Shamisen
12. Vibraphone	44. Contrabass	76. Pan Flute	108. Koto
13. Marimba	45. Tremolo Strings	77. Blown Bottle	109. Kalimba
14. Xylophone	46. Pizzicato Strings	78. Shakuhachi	110. Bagpipe
15. Tubular Bells	47. Orchestral Harp	79. Whistle	111. Fiddle
16. Dulcimer	48. Timpani	80. Ocarina	112. Shanai
17. Drawbar Organ	49. String Ensemble 1	81. Lead 1 (square)	113. Tinkle Bell
18. Percussive Organ	50. String Ensemble 2	82. Lead 2 (sawtooth)	114. Agogo
19. Rock Organ	51. Synth Strings 1	83. Lead 3 (calliope)	115. Steel Drums
20. Church Organ	52. Synth Strings 2	84. Lead 4 (chiff)	116. Woodblock
21. Reed Organ	53. Choir Aahs	85. Lead 5 (charang)	117. Taiko Drum
22. Accordion	54. Voice Oohs	86. Lead 6 (voice)	118. Melodic Tom
23. Harmonica	55. Synth Voice	87. Lead 7 (fifths)	119. Synth Drum
24. Tango Accordion	56. Orchestra Hit	88. Lead 8 (bass + lead)	120. Reverse Cymbal
25. Acoustic Guitar (nylon)	57. Trumpet	89. Pad 1 (new age)	121. Guitar Fret Noise
26. Acoustic Guitar (steel)	58. Trombone	90. Pad 2 (warm)	122. Breath Noise
27. Electric Guitar (jazz)	59. Tuba	91. Pad 3 (polysynth)	123. Seashore
28. Electric Guitar (clean)	60. Muted Trumpet	92. Pad 4 (choir)	124. Bird Tweet
29. Electric Guitar (muted)	61. French Horn	93. Pad 5 (bowed)	125. Telephone Ring
30. Overdriven Guitar	62. Brass Section	94. Pad 6 (metallic)	126. Helicopter
31. Distortion Guitar	63. Synth Brass 1	95. Pad 7 (halo)	127. Applause
32. Guitar harmonics	64. Synth Brass 2	96. Pad 8 (sweep)	128. Gunshot

Table 9.3 GM Percussion Patch Map for Channel 10

35. Acoustic Bass Drum	51. Ride Cymbal 1	67. High Agogo
36. Bass Drum 1	52. Chinese Cymbal	68. Low Agogo
37. Side Stick	53. Ride Bell	69. Cabasa
38. Acoustic Snare	54. Tambourine	70. Maracas
39. Hand Clap	55. Splash Cymbal	71. Short Whistle
40. Electric Snare	56. Cowbell	72. Long Whistle
41. Low Floor Tom	57. Crash Cymbal 2	73. Short Guiro
42. Closed Hi Hat	58. Vibraslap	74. Long Guiro
43. High Floor Tom	59. Ride Cymbal 2	75. Claves
44. Pedal Hi-Hat	60. Hi Bongo	76. Hi Wood Block
45. Low Tom	61. Low Bongo	77. Low Wood Block
46. Open Hi-Hat	62. Mute Hi Conga	78. Mute Cuica
47. Low-Mid Tom	63. Open Hi Conga	79. Open Cuica
48. Hi Mid Tom	64. Low Conga	80. Mute Triangle
49. Crash Cymbal 1	65. High Timbale	81. Open Triangle
50. High Tom	66. Low Timbale	

Note: The numbers represent the keynotes on a MIDI keyboard, which are followed by the percussion instrument's name.

Table 9.4 details the differences between file sizes as they range from voice-quality, 8-bit, 11-kHz files, all the way to the professional 16-bit, 44.1-kHz rates that are used to encode sounds with CD quality.

Sound File Formats

Although several formats exist for saving sample file and sound file data onto computer-based storage media, only a few formats have been universally adopted by the industry. These standardized formats make it easier to exchange files between compatible sample editor programs.

By far, the most common file type is the Wave (or .WAV) format, which is used with Windows and most DOS-based systems. This file type can be used to save waveform data at any bit rate and sample rate supported by your soundcard or system. Another file format, commonly used with Mac computers, is the Audio Interchange File Format (or .AIF). As with Wave files, .AIF files can encode both monaural and stereo sampled sounds with a variety of sample rates and bit rates. Another format used

Table 9.4 Digital Audio File Size Comparisons

Sample Rate (kHz)	Bit Structure	Bytes/minute
11	8-bit/mono	660 KB/minute
11	8-bit/stereo	1.3 MB/minute
11	16-bit/stereo	1.3 MB/minute
11	16-bit/stereo	2.6 MB/minute
22	8-bit/mono	1.3 MB/minute
22	8-bit/stereo	2.6 MB/minute
22	16-bit/mono	2.6 MB/minute
22	16-bit/stereo	5.3 MB/minute
44.1	8-bit/mono	2.6 MB/minute
44.1	8-bit/stereo	5.3 MB/minute
44.1	16-bit/mono	5.3 MB/minute
44.1	16-bit/stereo	10.5 MB/minute

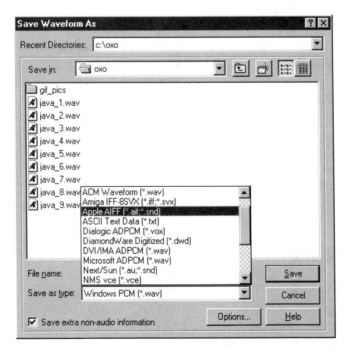

Figure 9.6 Digital editors can often save waveform data in any number of formats. (Courtesy of Syntrillium Software Corporation, www.syntrillium.com)

in media production is the Sound Designer II or .SND format, which was developed by Digidesign for their Pro-Tools and other digital editor systems.

Due to the wide range of available sound file formats, most professional editors are capable of both reading and saving files in a number of cross-platform formats and in a wide range of bit and sample rate structures (Figure 9.6).

Multimedia and the Web

Surfin' on the Internet. . . . This household phrase has become synonymous with jumping onto the web, browsing the sites, and grabbin' onto all of those hot digital graphics, Wave files, and videos that might wash your way.

Although this analogy often holds true for text-based information, many of you know that once you get out into deeper waters where the graphic and audio "waves" become much larger in size, pulling in the big ones often takes so much time that you're often left treading water, which leads us to. . . .

Streaming Audio

Let's face it, folks, even though the Internet has profoundly changed many of our lives, the relatively slow modem transfer speeds of 14.4 and 28.8 kbps that occurred during the infancy years of the "web" had taken us back to the premultimedia days when graphics were sluggish and real-time video was but a dream.

Although data speeds have increased over the years, the "bottleneck" of the web has forced hardware and programming designers to be innovative in coming up with ways to improve data transmission speeds. Regardless of the media type, these solutions often takes any of three forms: reducing the bit rate, reducing the sample rate, or employing compression techniques.

Even with compression, we generally accept that the web can't transmit .WAV or .AIF sound files in real time with any sense of stability. The only option, without special hardware and additional phone lines, is to download an entire file and then play it back from the hard drive (a process that could take several minutes to several hours, even at 56k!). However, through the use of special compression techniques, the dilemma of "streaming" sound over the Internet in real time has been tackled by a small number of companies, allowing audio to finally become an integral part of the net's multimedia experience.

Figure 9.7 RealAudio can be used to play audio off the Internet in real time. (Courtesy of Progressive Networks, Inc., www.real.com)

Most of the real-time audio glory goes to a Seattle-based company called Progressive Networks, with the introduction of their server/player application called RealAudio (Figure 9.7). In fact, with free distribution of millions of player applications to Internet sites worldwide since its introduction in 1995, RealAudio has pretty much become the *de facto* standard for streaming real-time audio over the web.

From a technical point of view, there's nothing magical about RealAudio, except the fact that it works—and works well. To begin with, RealAudio data is transmitted using any of more than 12 proprietary codes (called *codecs*) that range from transmission rates of 8 kbps (mono voice quality over RealAudio 1.0 with a frequency bandwidth of 8 kHz) all the way to 40 kbps (stereo music over a 16-kHz range) or 80 kbps (mono music over a 44-kHz range).

Although there are several codecs to choose from, the most common types compress audio into a "voice mode" format that transmits data over 14.4-KB lines in a way that's optimized for human speech. The second "Music Mode" compresses data in a way that is less harsh and introduces fewer artifacts over a greater dynamic range, thereby creating an algorithm that can more faithfully reproduce music with "near-FM" quality over 28.8-KB or faster lines. At the originating Internet site, the RealAudio server can automatically recognize which modem speed is currently in use and then transmit the data in the best possible audio format. This reduced data throughput ultimately means that the RealAudio player will take up very little of your computer's resources, allowing you to keep on working while audio is being played.

On a personal note, I'm a real fan of Real Audio's TimeCast service (www.timecast.com). This service gives you access to radio stations, live performances, and even video feeds from throughout the world. Several years ago, I threw a birthday party with a Mexican theme. Thanks to Real Audio, music was piped in live from my favorite radio station out of Mexico City. Olé!

Other forms of real-time and non-real-time streaming are also available on the web. The most notable of the real-time streamers is Shockwave from Macromedia (www.shockwave.com) and Liquid Audio (www.liquidaudio.com). Another media form that's gaining popularity is the MPEG 3 (.MP3) format. This standard does a really good job of compressing audio and sounds surprisingly good. Its current downfall (which might change by the time you read this . . . it's tough to read the crystal ball of technology) is that most MP3 players don't stream audio, and if they do, they use incompatible ways to achieve it.

Streaming Video

In addition to real-time audio, video can also be streamed over the Internet using various forms of codecs. Again, the most accepted provider for delivering video over the Internet is RealVideo from Progressive Networks. RealVideo delivers newscast-quality video over 28.8-kbps modems, full-motion-quality video using V.56 (56-kbps) and ISDN (56/64-kbps) modems, and near TV broadcast quality video at LAN rates or "broadband" speeds (100 kbps and above). On the client side, RealVideo delivers easy-to-use interactive features, such as video seeking and scanning, clickable video regions, and "buffered play" for greater video quality using slower 28.8-kbps modems. In addition to the RealVideo player, Progressive Networks offers authoring tools, such as RealPublisher, to create audio and video content using encoding wizards or templates. It can also be used to create compressed audio and video content from existing .WAV, .AU, .AVI, and Quicktime files. Using such a package, you can integrate audio and video media into your e-mail, allowing you to hear and see e-mail clips, which can be sent as an attachment or as a URL pointer to your web pages.

10

SYNCHRONIZATION

Over the years, electronic music has evolved into an indispensable production tool within the fields of professional audio, video, and film production. Audio transports, video decks, and electronic instruments routinely operate in computer-controlled tandem, allowing production teams to create today's records, video, and film soundtracks and multimedia scores.

One of the central concepts behind electronic music production is the ability of sequencers, audio and video transports, musical instruments,

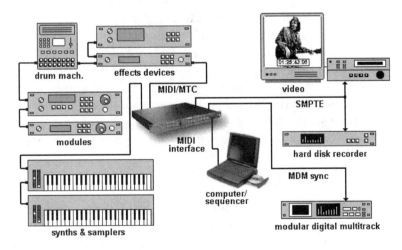

Figure 10.1   Example of a synchronized audio production system.

digital audio workstations, controllers, and effects to communicate and interface directly with each other in a real-time environment (Figure 10.1). The method that makes it possible for these devices and multiple media to maintain a relative time relationship is known as *synchronization* (often abbreviated as *sync*).

Synchronization is achieved whenever two or more related events occur at precisely the same point in time. With respect to analog audio and video systems, this is achieved by interlocking the transport speeds of two or more machines. Within computer-related systems (such as digital audio and MIDI), sync is commonly maintained through the use of an internal or external clock-timing pulse that's directly imbedded within the system's digital word structure (e.g., AES/EBU, MIDI sample dump, and MIDI). Because sync usually needs to be maintained between mechanical and digital devices, modern technology has come up with some rather ingenious forms of system communication and data translation so that dissimilar devices can talk to each other. In this chapter, we look at the various forms of synchronization that are used with both analog devices and digital devices, and also look at current systems for maintaining sync between various media.

Synchronization between Analog Transports

The basic theory behind keeping analog tape transports in sync with each other is tied to the fact that all the transport speeds involved in the production process don't have to be constant, they just have to maintain the same relative speed. That is to say, if the speed of one device changes, other devices must correspondingly change their speed in order to maintain sync.

Analog tape devices (unlike most of their digital counterparts) have a difficult time maintaining a perfectly constant tape speed. For this reason, sync between two or more machines would be impossible over any reasonable program length, as any time relationship would soon be lost as a result of such factors as voltage fluctuations and tape slippage. Thus, a means of interlocking these devices within a production system is essential.

SMPTE Time Code

The standard method for locking audio, video, and most other professional media together is *SMPTE (Society of Motion Picture and Television Engineers) time code*. Time code is used to identify an exact location on a magnetic tape by sequentially assigning a digital address at specific points over the

Figure 10.2
Readout of a SMPTE time-code address.

hour   minutes   seconds   frames

duration of a program. These identifying addresses can't slip and always retain their original-location identity, allowing for the continual monitoring of a tape's position to an accuracy of 1/30th of a second. The smallest and most accurate time points are called *frames* (a term taken from film production). Each audio or video frame is tagged with a unique identifying number, known as a *time-code address*. This eight-digit address is displayed in the form 00:00:00:00, where the successive pair of digits represent Hours:Minutes:Seconds:Frames (Figure 10.2).

The recorded time-code address is used to locate a position on magnetic tape in much the same manner that a postal carrier uses an address to deliver the mail. For example, if a mail carrier is to deliver a letter to 2100 Hoover Road, he or she knows precisely where to deliver it because the house doesn't move from its assigned address number (Figure 10.3A). Similarly, a time-code address can be used to locate specific areas on a magnetic tape. For example, let's assume that we would like to lay down the sound of a squealing car at 00:12:53:18 onto a time-encoded multitrack tape that

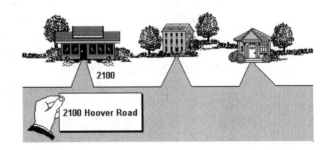

(A) Delivery of a letter to a specific postal address.

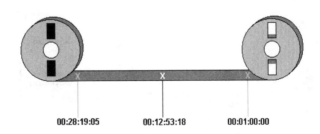

(B) A SMPTE time-code address on a magnetic tape.

Figure 10.3
Relative address locations.

00:28:19:05   00:12:53:18   00:01:00:00

begins at 00:01:00:00 and ends at 00:28:19:05 (Figure 10.3B). Through the monitoring of the address code (by means of a computer screen, tape position counter, or synchronizer), we can locate the position that corresponds to this tape address and insert the effect at that time . . . "SCREEEE!"

Time-Code Word

The time-encoded information that's recorded within each audio or video frame is known as a *time-code word*. Each word is divided into 80 equal segments called *bits*, which are numbered consecutively from 0 to 79. One word occupies an entire audio or video frame, so that for every frame there is a corresponding time-code address. Address information is contained in the digital word as a series of binary 1's and 0's. These bits are electronically generated as fluctuations (or shifts) in the voltage level of the time code's data signal. This method of encoding serial information is known as *biphase modulation* (Figure 10.4). When the recorded biphase signal shifts either up or down at the extreme edges of a clock period, the pulse is coded as a binary 0. A binary 1 is coded for a bit whose signal pulse shifts halfway through a clock period. The advantage of this encoding method is that detection relies on shifts within the pulse and not on the pulse's polarity. This means that the time code can be read in either the forward or reverse direction, as well as at fast or slow shuttle speeds.

The 80-bit time-code word (Figure 10.5) is further subdivided into groups of four bits, with each group representing a specific coded piece of information. Each of these 4-bit segments is encoded as a representation of a decimal number ranging from 0 to 9, written in binary-coded decimal (BCD) notation. When a time-code reader detects the pattern of 1's and 0's in a 4-bit group, it interprets the information as a single decimal number. Eight of these 4-bit groupings combine to constitute an address in hours, minutes, seconds, and frames.

The 26 digital bits that make up the time-code address are joined by an additional 32 bits called *user bits*. This additional set of encoded information (which is also represented as an 8-digit number) is set aside so that high-end production users can enter personal ID information. The SMPTE Standards

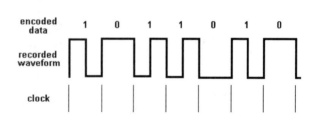

Figure 10.4
Biphase modulation encoding.

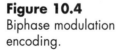

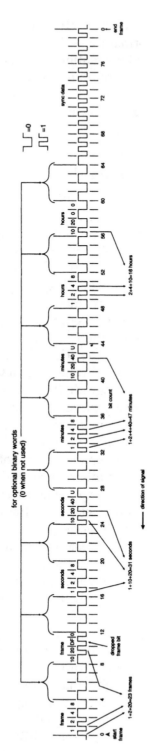

Figure 10.5 Biphase representation of the SMPTE/EBU time-code word.

Committee has placed no restrictions on the use of this "slate code," which can contain such information as date of shooting, take ID, reel number, etc.

Another form of information that's encoded into the time-code word is *sync data*. Sync data is found in the final 16 bits of a time-code word and is used to define the end of each frame. Because time code can be read in either direction, sync data also signals the controlling device as to which direction the tape or device is moving.

Time-Code Frame Standards

In productions using time code, it's often important for the readout display to be directly related to the actual elapsed program time, particularly when dealing with the exacting time requirements of broadcasting. In the case of the original black-and-white (monochrome) video signal, a frame rate of exactly 30 frames/sec is used. If this time code (commonly known as *nondrop code*) is read, the time-code display, program length, and actual clock on the wall will all be in agreement.

This simplicity was broken, however, when the National Television Standards Committee set the frame rate for the color video signal at approximately 29.97 frames/sec. This means that when a time-code reader that's set to read the monochrome rate of 30 frames/sec is used to read a 29.97 frames/sec color program, the time-code readout would pick up an extra 0.03 frames for every passing second (30 − 29.97 = 0.03 frames/sec). Over the duration of an hour, the address readout will differ from the actual clock by a total of 108 frames or 3.6 seconds.

To correct for this discrepancy, a form of frame adjustment is carried out by dropping 108 frame numbers over the course of an hour. This type of code used for color has come to be known as *drop-frame code*.

In correcting this timing error, two frame counts for every minute of operation are omitted from the code, with the exception of minutes 00, 10, 20, 30, 40, and 50. This has the effect of adjusting the frame count to agree with the actual elapsed program duration.

In addition to the color 29.97 drop-frame code, a 29.97 nondrop frame color standard is also commonly used in video production. When using this *nondrop time code*, the frame count will always advance one count per frame, with no drops in count. Although this will again result in a disagreement between the frame count and the actual program time, this method has the distinct advantage of easing time calculations during the video editing process. This is due to the fact that frame drops don't need to be taken into account when adding or subtracting nondrop SMPTE addresses.

Another frame rate format that's used throughout Europe is *EBU (European Broadcast Union) time code*. EBU utilizes the same 80-bit SMPTE encoded word. However, it differs in that it uses a 25 frames/sec frame rate that runs at exactly 24 frames/sec for both monochrome and color video signals.

Film media, on the other hand, uses a standardized 24 frames/sec format that differs from SMPTE time code. Despite this, many newer synchronization and digital audio devices can communicate with film-related systems using the language of SMPTE.

LTC and VITC Time Code

Two methods are currently used for encoding time code onto magnetic tape for broadcast and production use: LTC and VITC. Time code that's recorded onto an audio track or cue track of a video deck is known as *longitudinal time code (LTC)*. This common system encodes a biphase time-code signal onto an analog audio or cue track as a modulated square-wave signal at a bit rate of 2400 bps.

The recording of a perfect square wave onto a magnetic audio track is difficult, even under the best of conditions. For this reason, SMPTE has set forth a standard allowable rise time of 25 +/- 5 microseconds for the recording and reproduction of code. This is equivalent to a signal bandwidth of 15 kHz, which is well within the range of most professional audio recording devices.

Variable-speed time-code readers can commonly decode time-code information at shuttle rates that range from 1/10th to 100 times normal playing speed. This is effective for most audio applications. However, in video postproduction, it's often necessary to view a videotape at slow or still speeds. Because LTC can't be read at speeds slower than 1/10th to 1/20th normal playing speed, two alternate methods must be used. One method is to "burn" the time-code address into the video image of a copy work tape, which uses a character generator to superimpose the corresponding address within an on-screen window (Figure 10.6). This visible "window dub" lets you easily identify exact address times, even at very slow or still picture shuttle speeds.

Another method used by major production houses is to stripe the picture with *VITC (Vertical Interval Time Code)*. VITC uses the same SMPTE address and user-code structure as does LTC. However, it's encoded onto videotape in an entirely different way. This method actually encodes the time-code information within the video signal itself, inside a field that's located outside of the visible picture scan area known as the *vertical blanking interval*. Because the time-code information is encoded into the video signal, it's possible for broadcast-quality, helical scan video recorders to read time code at extremely slow speeds and even still frame. Because time code can

Figure 10.6
Video image showing
burned-in time-code window.

now be accurately read at all speeds, this added convenience opens up an additional track on a video recorder for audio or cue information and eliminates the need for a window dub.

In most situations LTC code is preferred by electronic music production and standard video production houses because it's a more accessible and cost-effective protocol.

Jam Sync/Restriping Time Code

As we've seen, LTC operates by recording a series of square-wave pulses onto magnetic tape. Unfortunately, it's not a simple matter to record the square waves without moderate to severe waveform distortion. Although time-code readers are designed to be relatively tolerant of waveform amplitude fluctuations, the situation is severely compounded when the code is dubbed (copied) by one or more generations. For this reason a special feature, known as *jam sync*, has been built into most time-code synchronizers and MIDI interfaces that have extensive sync facilities.

It's basically the function of the jam sync process to read the time-code address numbers off of the original tape and regenerate fresh code onto the newly created copy during the dubbing stage or, if necessary, reconstruct missing or defective sections of code (Figure 10.7).

Currently, two common forms of jam sync are in use: one-time jam sync and continuous jam sync. *One-time jam sync* refers to a mode whereby, upon receiving a valid time code, the generator's output is initialized to the first valid address number that's received and begins to count in an ascending order on its own in a freewheeling fashion. Any deterioration or discontinuities in code are ignored, and the generator will produce fresh uninterrupted address numbers.

Continuous jam sync is used in cases where the original address numbers must remain intact and not be regenerated into a contiguous address count. Once the reader is activated, the generator will update the address count and each frame will be regenerated in accordance with the incoming address.

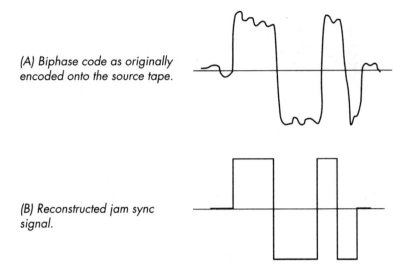

(A) Biphase code as originally encoded onto the source tape.

(B) Reconstructed jam sync signal.

Figure 10.7   Representation of the recorded biphase signal.

Distribution of LTC Signal Lines

Longitudinal SMPTE time code can be distributed throughout the production and postproduction system in a fashion that's similar to any other audio signal. It can be routed through any audio path or patch point via normal two-conductor shielded cables; and, since the time-code signal is biphase or symmetrical in nature, it is immune to polarity problems.

Time-Code Levels

One problem that can potentially plague systems using time code is crosstalk. This nuisance arises from high-level time-code signals that can "bleed" onto adjacent audio signals or recorded tape tracks, and be heard as a gruff, high-pitched warbling sound. Currently, no industry standard

Table 10-1   Optimum Time-Code Recording Levels

Tape Format	Track Format	Optimum Rec. Level
ATR	Edge track (highest number)	–5 VU to –10 VU
3/4-inch VTR	Audio 1 (L) track or time	–5 VU to 0 VU code track
1-inch VTR	Cue track or audio 3	–5 VU to –10 VU

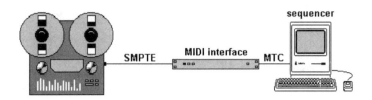

Figure 10.8   Example of a SMPTE-to-MIDI converter.

levels exist for the recording of time code onto magnetic tape. However, the levels shown in Table 10.1 have proven over time to give the best results and to reduce crosstalk

SMPTE to MIDI Conversion

A SMPTE-to-MIDI converter (Figure 10.8) serves to read SMPTE time code and convert it into MIDI Time Code. These converters are often designed into newer multiport MIDI interfaces. However, if you don't have such an interface, a few self-contained, stand-alone devices are on the market that can convert SMPTE to such MIDI-based sync protocols as SPP, direct time lock, and MTC. In certain cases these converters have been directly integrated into such non-MIDI devices as audio recorders and intelligent remote controllers. In most cases, these converters are able to output an LTC SMPTE signal, allowing an analog or video sync track to be "striped" with an ascending SMPTE code.

Synchronization in Electronic Music Production

The acceptance of MIDI and digital audio by the various production media has created the need for routinely synchronizing devices such as MIDI sequencers, digital audio editors, effects devices, and automated mixing systems. The following sections discuss the various types of sync that can be encountered in many of today's production environments.

Non-MIDI Synchronization

Several types of synchronization methods were used by older electronic instruments and devices designed before the MIDI specification was implemented. Although sync between these non-MIDI and MIDI instruments was a source of mild to major aggravation, many of these older devices can still be found in MIDI setups because of their distinctive and wonderful sounds.

Click Sync

Click sync or *click track* refers to the metronomic audio clicks that are generated to communicate tempo. These are produced once per beat or once per several beats (as occurs in cut time or compound meters).

Often, a click or metronome is designed into a MIDI interface or sequencing software to produce an audible tone or to trigger MIDI instrument notes that can be used as a tempo guide. These guide clicks help a musician keep in tempo with a sequenced composition.

Certain sync boxes and older drum machines can sync a sequence to a live or recorded click track. They can do this by determining the beat based on the tempo of the clicks and then output a MIDI start message (once a sufficient number of click pulses has been received for tempo calculation). A MIDI stop message might be transmitted by such a device whenever more than two clicks have been missed or whenever the tempo falls below 30 beats/minute.

Note that this sync method doesn't work well with rapid tempo changes, because chase resolutions are limited to one click per beat (1/24th the resolution of MIDI clock). Thus, it's best to use a click source that is relatively constant in tempo.

TTL and DIN Sync

One of the early, most common ways to lock sequencers, drum machines, and instruments together, before the adoption of MIDI, was TTL 5-volt sync. In this system, a musical beat is divided into a specific number of clock pulses per quarter note (PPQN), which varies from device to device. For example, *DIN sync* (a form of TTL sync, which is named after the now famous 5-pin DIN connector) transmitted at a rate of 24 PPQN.

TTL can be transmitted in either one of two ways. The first and simplest way uses a single conductor that passes a 5-volt clock signal. Quite simply, once the clock pulses are received by a slave device, it will start playing and synchronize to the incoming clock rate. Should these pulses stop, the devices will also stop and wait for the clock to again resume. The second method uses two conductors, both of which transmit 5-volt transitions. However, with the system, one line is used to constantly transmit timing information, while the other is used for start/stop information.

MIDI-Based Sync

In current MIDI production, the most commonly found form of synchronization uses the MIDI protocol itself for the transmission of sync messages. These messages are transmitted along with other MIDI data over standard MIDI cables, with no need for additional or special connections.

MIDI Sync

The most commonly used and most basic form of system synchronization is known as *MIDI Sync*. This "behind-the-scenes" protocol is used to help keep the system in a relative form of internal sync through the transmission of real-time MIDI messages over standard MIDI cables.

MIDI Real-Time Messages

MIDI real-time messages are made up of four basic types, which are each one byte in length: timing-clock, start, continue, and stop messages. The *timing-clock message* is transmitted to all devices within the MIDI system at a rate of 24 PPQN, although certain devices have been programmed to divide this structure into 24 clock signals per metronomic beat. This method is used to improve the system's timing resolution and simplify timing when working in nonstandard meters (i.e., 3/8, 5/16, 5/32).

The MIDI *start command* instructs all connected devices to start playing from the beginning of their internal sequences on receipt of a timing-clock message. Should a program be in midsequence, the start command will reposition the sequence at the beginning and begin to play.

On transmission of a MIDI *stop command* all of the connected devices will stop at their current position and wait for a message to follow. Following the receipt of a MIDI stop command, a MIDI *continue message* instructs all sequencers and/or drum machines to resume playing from the precise point at which the sequence was stopped. Certain older MIDI devices (most notably drum machines) aren't capable of sending or responding to continue commands. In such cases, the user must either restart the sequence from its beginning or manually position the device to the correct measure.

Song Position Pointer

In addition to MIDI real-time messages, the *song position pointer (SPP)* is a MIDI system common message that isn't commonly used in current-day production. Essentially, SPP keeps track of the current position by noting how many measures have passed since the beginning of a sequence. Each pointer is expressed as multiples of 6 timing-clock messages, and is equal to the value of a 16th note.

The song position pointer can synchronize a compatible sequencer or drum machine to an external source from any position within a song containing 1024 or fewer measures. Thus, when using SPP, it is possible for a sequencer to chase and lock to a multitrack tape from any measure in a song.

Within such a MIDI/tape setup, a specialized sync tone is transmitted that encodes the sequencer's SPP messages and timing data directly onto tape as a modulated signal. Unlike SMPTE time code, the encoding method isn't standardized between manufacturers. This lack of standardization could prevent SPP data written by one device from being decoded by another device that uses an incompatible proprietary sync format.

Unlike SMPTE, where tempos can be easily varied by inserting a tempo change at a specific SMPTE time, once the SPP control track is committed to tape, the tape and sequence are locked into this predetermined tempo or tempo map.

SPP messages are usually transmitted only while the MIDI system is in the stop mode, in advance of other timing and MIDI continue messages. This is due to the relatively short time period that's needed to locate the slaved device to the correct measure position.

FSK

In the pre-MIDI days of electronic music, musicians discovered that it was possible to sync instruments that used TTL 5-volt sync to a multitrack tape recorder. This was done by recording a master sync square-wave pulse onto tape (Figure 10.9A). Since the most common pulse in use was 24 and 48 PPQN, the recorded square wave consisted of an alternating 24- or 48-Hz signal.

Although this system worked, it wasn't without its difficulties, because the synchronized devices relied on the integrity of the square wave's sharp transition edges to provide the clock. Because tape is notoriously bad at reproducing a square wave (Figure 10.9B), the poor frequency response and reduced reliability at low frequencies mandated that a better system for synchronizing MIDI to tape be found. The initial answer was in *frequency shift keying*, better known as *FSK*.

FSK works in much the same way as the TTL sync track. However, instead of recording a low-frequency square wave onto tape, FSK uses two, high-frequency square-wave signals for marking clock transitions (Figure 10.9C). In the case of the MPU-401/compatible interface, these two frequencies are 1.25 and 2.5 kHz. The rate at which these pitches alternate determines the master timing clock to which all slaved devices are synched. These devices are able to detect a change in modulation, convert these into a clock pulse, and advance their own clocks accordingly.

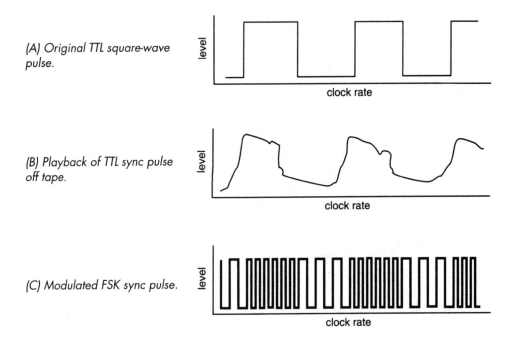

(A) Original TTL square-wave pulse.

(B) Playback of TTL sync pulse off tape.

(C) Modulated FSK sync pulse.

Figure 10.9   TTL and FSK sync waveforms.

Unlike most other forms of sync, FSK triggers and plays the sequence relative to the initial clock that's recorded onto tape. As such, the sequence must be positioned at its beginning and the tape "must" be cued to a point before the beginning of the song. Once the initial sync pulse is received, the sequencer will begin playback. Should a mistake happen, you'll need to recue the song back to its beginning points. . . . Now you can see why this form of sync died out.

MIDI Time Code

For decades, the SMPTE time code has been the standard timing reference in audio and video production. In order for MIDI-based devices to synchronize to this absolute timing reference with little or no additional hardware, *MIDI time code* or *MTC* was developed as a cost-effective and easily implemented means for translating SMPTE time code into MIDI messages that can be easily distributed throughout the MIDI chain.

MTC doesn't replace MIDI 1.0; rather, it's an extension of MIDI since it uses existing message types that were either previously undefined or were being used for other, nonconflicting purposes, such as the sample-dump standard. In practice, MTC uses a reasonably small percentage of a MIDI line's available bandwidth (about 7.68% at 30 frames/sec). For this reason, it's often a good idea to keep the MTC signal path separate from the MIDI performance paths, so as to reduce the possibility of data overloading or delay. This is not a hard and fast rule and will depend on your specific applications, as well as your ears.

MIDI Time-Code Control Structure

The MIDI time-code format can be broken into two parts: time code and MIDI cueing. The time-code capabilities of MTC are relatively straightforward and allow both MIDI and non-MIDI devices to attain synchronous lock or to be triggered via SMPTE time code. MIDI cueing is a format that informs MIDI devices of events that are to be performed at a specific time (such as load, play, stop, punch-in, punch-out, and reset). This language can be used to communicate commands to intelligent MIDI devices that can prepare for a specific event in advance, and then execute a command on cue.

MIDI Time-Code Commands

MIDI time code uses three message types: quarter-frame messages, full messages, and MIDI cueing messages. These are discussed in the following subsections.

Quarter-Frame Messages
Quarter-frame messages are transmitted only while the system is running in real or variable-speed time, and in either the forward or reverse directions. In addition to providing the system with its basic timing pulse, four frames are generated for every SMPTE time-code field. This means that should you decide to use non-drop frame code (30 frames/sec), the system would transmit 120 quarter-frame messages per second.

Quarter-frames messages should be thought of as groups of eight messages, which encode the SMPTE time in hours, minutes, seconds, and frames. Since eight quarter frames are needed for a complete time-code message, SMPTE time is updated every two frames. Each quarter-frame message contains 2 bytes; the first being Fl (the quarter-frame common header), and the second byte contains a nibble (4 bits) that represents the message number 0–7, and a nibble for each of the digits in a time field (hours, minutes, seconds, or frames).

Full Messages
Quarter-frame messages aren't sent while in the fast forward, rewind, or locate modes, because they would unnecessarily clog or outrun the MIDI

data lines. When in any of these shuttle modes, a full message (which encodes the complete time-code address within a single message) is used.

Once a fast shuttle mode is entered, the system will generate a full message and then place itself into a pause mode until the time-encoded device has arrived at its destination. After the device has resumed playing, MTC will again begin sending quarter-frame messages.

MIDI Cueing Messages

MIDI cueing messages are designed to address individual devices or programs within a system. These 13-bit messages can be used to compile a cue or edit decision list, which in turn instructs one or more devices to play, punch in, load, stop, etc., at a specific time. Each instruction in a cueing message contains a unique number, time, name, type, and space for additional information. At present only a small percentage of the possible 128 cueing-event types have been defined.

Digital Audio Synchronization

Coverage of synchronization would be incomplete without discussing it in the context of digital audio and hard disk based systems. Because digital audio is an important part of modern-day audio and audio-for-visual production, an understanding of digital sync becomes important when you're working in an environment where digital audio devices are to be synchronized to each other or with video and analog media.

The Need for a Stable Timing Reference

The process of maintaining a synchronous lock between digital audio devices or between digital and analog systems differs fundamentally from the process that's used to maintain relative speed between analog transports. This difference is due to the fact that a digital system can achieve synchronous lock by adjusting its playback sample rate (and thus its speed and pitch ratio) to match precisely the relative playback speed of the master transport.

Whenever a digital system is synchronized to a time-encoded master source, the need for a stable timing source is extremely important. Such an accurate timing reference may be required to keep jitter (in this case, an increased distortion due to rapid pitch shifts) to a minimum. In other words, the program speed of the source should vary as little as possible over time in order to prevent any adverse effects in the digital signal's quality. For example, all analog tape machines exhibit speed variations caused by tape slippage and transport irregularities (a basic fact of analog life known as *wow* and *flutter*). If you were to closely synchronize a disk-based recorder to a time-encoded analog source that contains excessive wow and flutter, the digital system would be constantly called on to speed up and slow down to precisely match these speed fluctuations. The best way to avoid these problems would be to use a transport type that's more stable, such as a digital audio or video recording system.

Black Burst

Whenever a video signal is copied from one machine to another, it's essential that the scanned data (containing timing, video, and user information) be copied in perfect sync from one frame to the next. Failure to do this results in severe picture breakup or, at best, the vertical rolling of a black line over the visible picture area.

Copying video from one machine to another generally isn't a problem because the VCR or VTR that's doing the copying is able to obtain its sync source from the playback machine without a hitch. A video postproduction house, however, often uses multiple video decks, switchers, and edit controllers in the production of a single program. Mixing and switching between these sources without a stable sync source will often result in various forms of video breakup and chaos.

This sync nightmare can easily be solved by using a single timing source known as a *black burst generator*. This generator produces an extremely stable timing reference (called *black burst* or *house sync*) that has a clock frequency of exactly 15,734.2657 Hz. This signal is used to synchronize the video frames and time code addresses that are received or transmitted by every video-related device in a production facility so that the frame and address's leading edge occurs at exactly the same instant in time (Figure 10.10).

By resolving all video and audio devices to a single black burst reference, you're assured that the relative frame transitions and speeds

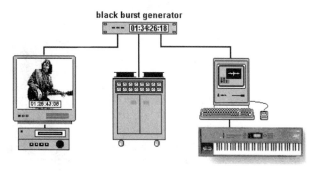

Figure 10.10
Example of a system whose overall timing elements are locked to a black burst reference signal.

throughout the system will be consistent and stable. This holds true even for analog machines because their transports can be locked (in a slave fashion) to this reference, thereby smoothing any inherent wow and flutter.

Synchronization Methods

Digital audio can obtain its clocking source from one of two places: an internal source or an external source. Whenever a digital device is recording audio as a stand-alone machine, the system's own internal quartz crystal oscillator serves as the clocking source. However, when two machines are used to make a digital copy, the device doing the recording derives its clock pulse from the playback machine's internal timing circuitry (Figure 10.11). In the case of the AES/EBU and S/PDIF digital transmission formats, this clock is embedded within the bit stream itself and doesn't require additional connections to be made.

Theoretically, when a digital device is synchronized to an external timing reference, its sample rate must be altered to match variations in that reference. In actual practice, however, digital audio systems can synchronize to various audio or visual media in a number of ways, depending on the actual application and the degree of timing accuracy that's required. These synchronization types are wild (no sync), SMPTE trigger, and continuous sync.

Wild Sync

When two signals are played back without any form of synchronization, the signals are said to be *wild* or *on the fly*, meaning that no form of sync is used at all (Figure 10.12). Under these conditions, the various transports can be manually triggered and any digital device's timing reference will be derived solely from its internal clock. For example, a CD player can be used to insert a non-time-dependent music track into a video, or it could be used to insert a wild sound effect cue. Just press play at the right time and . . . BLAM!

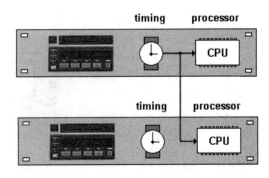

Figure 10.11
Whenever a signal is digitally copied, the device that's recording will derive its timing reference from the playback machine's clock pulse.

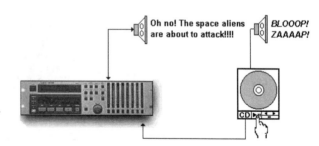

Figure 10.12
A wild signal is manually triggered without any reference to a time-encoded source.

SMPTE Trigger Sync

SMPTE trigger sync is often used as a straightforward way to achieve sync in most forms of digital media production. Trigger sync generally works by placing any number of sound or event cues into a playlist. Such a playlist can be assembled from a MIDI sequencing program (in order to trigger sound cue events from a sampler), from a digital audio editing program (triggering hard disk sound file cues), or even from a video or other edit system for triggering external devices (such as a professional CD player).

In these cases, once an event has been triggered, the digital device will begin to play the event in a wild, nonsynchronous fashion. Once playback has begun, the timing will be relatively stable as it is referenced to the digital device's own internal clock (Figure 10.13). Over longer periods of time, this form of relative sync might not be stable enough to lock an on-screen dialog segment that lasts over a few minutes in length. Under such conditions, a more precise form of synchronization is necessary.

Continuous SMPTE Sync

If longer segments are to be synched to a time-encoded source, or if a source's timing element is unstable, a method known as *continuous SMPTE sync* should be used in order to keep the two systems in close sync. When using continuous sync (Figure 10.14), once an audio file or segment has begun to playback audio, it will begin to read and resolve its timing reference to the

Figure 10.13
Once a wild trigger event has occurred, the digital device plays back the event using its own internal clock as a reference.

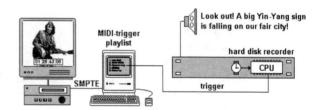

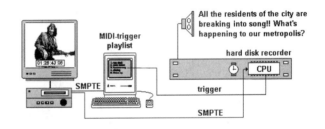

All the residents of the city are breaking into song!! What's happening to our metropolis?

MIDI-trigger playlist

hard disk recorder

CPU

SMPTE

trigger

SMPTE

Figure 10.14
Example of a system working in the continuous sync mode.

incoming time code that's being supplied from the master device. In this way, the slaved digital audio system can quickly chase to the proper SMPTE location point; thereafter, it will remain in tight sync by varying its sample rate to match precisely the time code's speed.

Keep in mind that an unstable master timing reference can cause the audio quality to be degraded (due to excessive sample rate fluctuations or jitter). You might wish to remedy this by locking both systems to a stable master timing source, such as black burst, or by turning off the continuous sync function, thereby playing back audio in a standard SMPTE trigger fashion.

Proprietary Sync Methods for Modular Digital Multitrack Recorders

Modular digital multitrack recorders, such as the Tascam DA-98 and Alesis ADAT, encode a proprietary form of time-encoded sync data directly in the data stream along with audio information. This sync pulse can be used to synchronously lock more than one MDM together to give you access to more tracks.

MDMs can also be locked to an external device or computer-based application using more traditional forms of sync (such as SMPTE, MTC, or black burst). The simplest of these uses a simple, cost-effective interface (Figure 10.15) that can translate an MDM's sync code into MTC or generated LTC SMPTE (and vice versa). In this way, a digital multitrack can be easily rigged so as to control other slave devices, such as a sequencer or hard disk recorder.

Figure 10.15
JLCooper's dataSYNC2 MIDI synchronizer for the Alesis ADAT. (Courtesy of JLCooper Electronics, www.jlcooper.com)

Other, more sophisticated interface systems can link multiple MDMs directly to your computer (often locking them to a digital audio workstation), allowing all of the transport and remote track functions to be controlled directly from the computer. Other interface types offer additional sync and sample rate options that can be useful within a high-end audio or video production setting.

Real-World Sync Applications for Using Time Code and MIDI Time Code

Before we delve into the many possible ways that a system can be set up to work in a time-code environment, you need to understand that each system has its own particular "personality," and that the connections, software and operation of one system might differ greatly from that of another. This is often due to factors such as system complexity, installed hardware, and the type of software systems that are loaded into the computer.

Larger, more expensive setups that are used to create TV and film soundtracks will often involve extensive time-code and system interconnections that can get fairly complex. Fortunately, the use of MIDI time code has greatly reduced the cost and simplicity of connecting and controlling a synchronous production system down to levels that can be easily managed by most experienced and novice electronic musicians.

For the remainder of this chapter, we will look into some of the basic concepts and connections that can be used to get your system up and running. Beyond these basic concepts, the best way to get your particular system working smoothly is to consult your manual, seek help from an experienced friend, or call the tech department for the particular piece of hardware or software that is giving you and your system the willies.

Master/Slave Relationship

Because synchronization is based on the timing relationship between two or more devices, it follows that the easiest way to achieve sync is to have one or more devices (known as *slaves*) follow the relative movements of a single transport or device (known as a *master*). The basic rule to keep in mind is that there can be only one master in a connected system. However, any number of slaves can be set to follow the relative movements of a master transport or device (Figure 10.16).

Generally, the rule for deciding which device will be the master in a production system can best be determined by either of two answers to the following questions: (1) From a practical standpoint, which device will

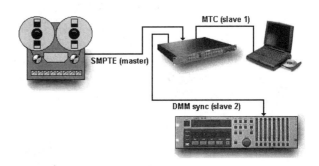

Figure 10.16
There can be only one master in a synchronized system, but there can be any number of slaves.

"want" to be master? (2) Which device will be the most stable master from a timing standpoint?

Audio Recorders
In many cases, whenever an analog tape recorder is connected in a time-code environment, this machine will want to act as the master, as a fair amount of costly hardware is required to "lock" an analog machine to an external time source. This is due to the fact that the machine's speed regulator (generally a DC capstan servo) must be connected in a feedback loop that compares its present location with that of the actual SMPTE location. As a result, it would be better for other, nonvideo devices (MIDI, digital audio editors, etc.) be slaved to this source.

The course of action is to "stripe" the highest track on a clean tape with SMPTE time code and then plug this track to the SMPTE IN on your MIDI interface (Figure 10.17). If you don't have a multiport interface or if your interface doesn't have an SMPTE input, you'll need to get hold of a box that converts SMPTE to MTC and then plug that into a MIDI in port. From here on out, it's smooth sailing. Simply select the MIDI interface's SMPTE sync driver or selected MIDI in port (if you need an MTC converter box) to be your sequencer or editor's sync source and you're in business.

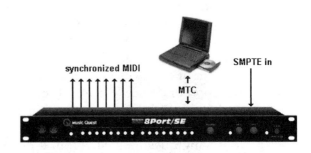

Figure 10.17
Routing the SMPTE track to the TC input on your interface can supply your system with MIDI time code

VCRs

Video is an extremely stable timing source. As a result, a video machine should almost invariably be a system master. In fact, without expensive hardware, a VCR can't easily be set to act as a slave, because a number of internal sync references would be thrown off and the picture would break up or begin to roll.

From a practical standpoint, locking other devices to a VCR is done in much the same way as with an analog tape machine. Professional video decks generally include a separate time code track (in addition to other tracks that are dedicated to audio) which must be striped with SMPTE.

Basically, the rule of thumb for time code is: If you're working on a project that was created "out-of-house," the videotape should be striped by the original production team. Striping your own code or erasing over their code with your own would render it useless, because it wouldn't relate to the original addresses or include any variations that might be a part of the original master source. In short, make sure that your working copy includes SMPTE that is a regenerated copy of the original code. Should you overlook this, you might run into timing and sync troubles later in the postproduction phase, when putting the music or dialog back together with the final video master.

MDMs

As a digital device, a modular digital multitrack machine is also an extremely stable timing reference and often works well as a master. Due to extensive sync and pitch shifting technology, these devices can also be slaves in a system without too much difficulty.

As with an analog machine, it's possible to record SMPTE onto the highest available track and then route this track to a valid SMPTE sync input. However, if you don't feel like losing a physical track to SMPTE, you might want to pick up an MDM sync interface that can translate the MDMs proprietary sync code into SMPTE or MTC (Figure 10.18). By plugging this box into your MIDI interface, you can select the interface's SMPTE driver as your sync source or you can plug the MDM interface's MTC output directly into a MIDI in port and select this port as the source for driving your sequencer and/or editor.

Software Applications

In general a MIDI sequencer will be programmed to act as a slave device. This is due to its ability to "chase" a master source by changing its timing clock. For example, an analog or video machine's SMPTE track can be plugged into the time-code input on a multiport or supporting MIDI interface and the sequencer's sync source can be easily derived from this source.

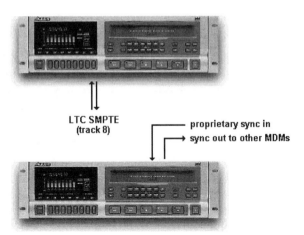

LTC SMPTE
(track 8)

proprietary sync in
sync out to other MDMs

Figure 10.18
Various sync interface sce-
narios when using modular
digital multitracks.

Digital audio editors, on the other hand, can often be set to act as a master or a slave. This will ultimately depend on the software because most professional editors can be set to chase (or be triggered to) a master source, however, not all of them can generate a sync signal (commonly MTC).

From a connections standpoint, most sequencer and editing packages will be flexible enough to let you choose from any number of available sync sources (regardless of whether they are connected to a hardware port, MIDI port, or virtual sync driver).

One of the cooler ways to lock a MIDI sequencer to digital audio is through the use of a *virtual MIDI router*. A VMR can be used to synchronize your system totally in the software domain, without the need for any external hardware or connections.

Currently, the only generic VMR that I'm aware of (which will work with most PC-based software systems) comes packaged with Sonic Foundry's Sound Forge editor. Version 4.0b and later of Sound Forge comes with a 32-bit version that works on Windows 95 and NT operating systems, while previous versions come bundled with a 16-bit version that works with Windows 3.x. It's also my understanding that Digidesign currently has a router that can virtually lock their Session PC editor to a MIDI sequencer. For the latest info, I'd suggest that you contact Sonic Foundry or your digital editor's manufacturer.

Although certain digital audio editors can't act as a master for generating time code in any form, there is often a simple solution. When faced with an all-slave software environment, you need a physical time-code master that can be routed to your editor and sequencer. This could be an MDM or analog recorder. But what if you simply want to sync the two pieces of software together without a tape machine? The only answer that I know of

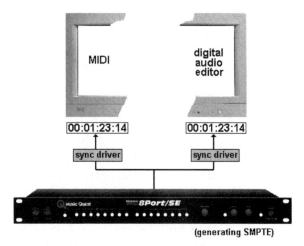

Figure 10.19
A multiport MIDI interface can lock devices and applications together by generating time code.

(generating SMPTE)

is to use an external device (such as a MIDI interface) that can generate time code (Figure 10.19).

Most modern multiport MIDI interfaces can generate and route SMPTE directly to receiving software in a virtual fashion. Often, when using such an interface, all you need to do is to select the interface's sync driver as your sync source for both the editor and sequencer. By simply pressing the "Generate SMPTE" button on the interface's front panel, the two programs will lock to the incoming master source, beginning at 00:00:00:00 or at any specified offset address. When using my Opcode 8-Port SE, I had to plug the SMPTE out jack to the SMPTE in jack. Once done, pressing the "SMPTE" button easily locked the two programs to the generated time code.

What if your interface doesn't have SMPTE capabilities? If you have a device or box around the room that can generate MIDI time code, try connecting its MTC out port to a MIDI in port on your interface. Sync can be achieved by routing the same MIDI in port to the sync in ports on both programs.

Frame Rates

When synchronizing devices, make sure that the rates of all of the slaves are set to the same frame setting as the master source. In the United States, these settings will most likely either be 30 or 29.97 frames/sec; 30 frames/sec (also known as *nondrop code*) is often used in music production that doesn't involve video production, whereas 29.97 frames/sec (commonly referred to as *drop code*) is often used in production work where color video is involved. I think it's safe to say that if you work in-house on projects that don't involve video, you're safe with 30 frames/sec. If you're working with a project that isn't in-house, *always* make sure that you know the original media's frame rate and stick with it. (When in doubt, always confer with the original production house, because mixed time code rates can cause timing problems down the road that can be a major pain to fix.)

SMPTE Offset Times

In the real world of audio production, songs don't always begin at 00:00:00:00. Let's say, that you were handed an ADAT tape that needed a synth track laid down onto track 7 of a song that goes from 00:11:24:03 to 00:16:09:21. Instead of inserting over 11 minutes of empty bars into your sequencer, you could simply open up your sync dialog box and create an *offset* of 00:11:24:03. Basically, this offset "slips" the relative times between the master and the slave (in this case, the slaved sequencer), so that bar 1 on the sequence will actually begin at 00:11:24:03. Both time-code addresses will agree with the incoming code, you'll simply be at the beginning of the sequence when the song starts. The same can also apply when syncing hard disk tracks to an external source that doesn't begin at 00:00:00:00.

An offset is also important when synchronizing devices to an analog or videotape source. As you probably know, it takes a bit of time for a tape transport to begin playing and to settle down (this time often quadruples whenever videotape is involved). If the time code were to begin at the head of the tape, it's very unlikely that you would want to start a program at 00:00:00:00, because of the time it takes the machine to get up to speed. As a result, you might want to offset the song, so that it starts at 00:00:30:00 or at 00:01:00:00.

As a closing remark, synchronization can be a simple procedure or it can be fairly complex, depending on your experience and the type of equipment that is involved. A number of books and articles have been written on the subject. To those who are serious about production, I'd suggest that you do your best to keep up on the subject. Although the fundamentals often stay the same, new technologies and techniques are constantly emerging. However, having said this, the best way to learn is simply by jumping in and doing it.

11

MIDI-BASED MIXING AND AUTOMATION

In the past, almost all commercial music was mixed by a professional recording engineer under the supervision of a producer and/or artist. With the emergence of electronic music and high-quality recording equipment, projects that have high production standards have become much more personal and cost effective in nature. This is largely due to the fact that MIDI and digital audio workspaces can now be owned by individuals, small businesses, and artists who take the time to become experienced in the commonsense rules of creative and commercial mixing.

Within the industry, it's a well-known fact that most professional mixers had to earn their "ears" by logging countless hours behind the console. Although there's no substitute for this expertise, the mixing abilities and ears of electronic musicians are also improving as equipment quality gets better and as they become more knowledgeable about proper mixing environments and techniques (often by mixing their own compositions).

Analog Mixers

Most analog audio production consoles used in professional recording studios are designed with similar controls and functional capabilities. They differ mostly in appearance, control location, on-board dynamic processing, signal routing capacities, and how they incorporate automation (if at all).

Before delving into the details of how a console works, you need to understand one of the most important concepts in all of audio technology: the *signal chain* (also known as the *signal path*). As is true with any audio system, the recording console can be broken down into separate signal paths. By identifying and examining the individual components that work together to form this chain, you can more easily understand the basic operations of any mixer or console, no matter how large or complex.

The trick to understanding these systems is to realize that each component has an input and an output, both of which are associated with an audio source and destination. In such a chain, the output of each source device must be connected to the input of the device that follows it until the end is reached. Whenever a link in this source-to-destination path is broken, no signal will pass.

Although this might seem to be a very simple concept, acknowledging it will save time and frustration when paths, devices, and spaghetti cables get out of hand. It's as basic as knit one, purl two: "An audio signal follows from one control or processing device to the next until the desired effect or goal is achieved."

The signal flow for each input of a modern console follows vertically down a plug-in strip known as an *I/O module* (Figure 11.1). *I/O* stands for input/output and is so named because all the associated electronics for a single track–channel combination are often located on a single circuit board. Because I/O module electronics are self-contained, they can be cost effectively fitted into a modular mainframe in a number of configurations.

Need More Inputs?

It's no secret that electronic music has had a major impact on the physical requirements of mixing hardware. This is largely due to the increased need for the large number of physical inputs, outputs, and effects that are commonly encountered in the modern project and MIDI production facility.

Although traditional analog mixer and recording console designs haven't significantly changed over the years, electronic music production has placed new demands on these devices. For example, newer project console designs that offer up to 32 and even 48 inputs have sprung up to meet the input demands that modern-day electronic instruments and digital editors place on them.

Whether you're using a larger console, or a mixer that has 16 or 24 inputs, it's easy to see how you might run out of input options when faced with synths that have 4 outputs, drum machines with 6, and samplers having up to 8 outputs. Although you might not plug all of these outputs into your mixer, you can still see how a system might easily outgrow a console's

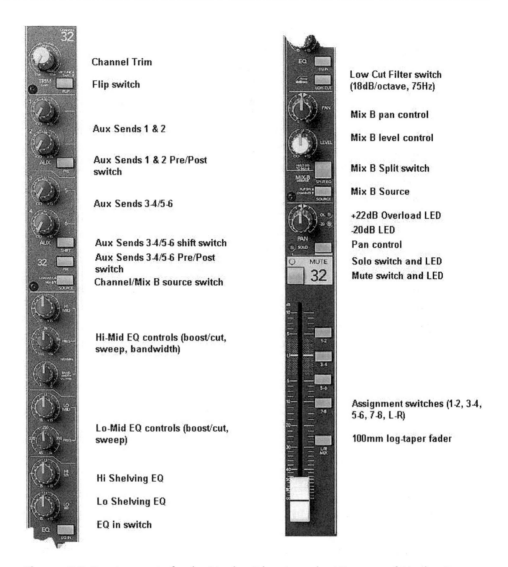

Channel Trim

Flip switch

Aux Sends 1 & 2

Aux Sends 1 & 2 Pre/Post switch

Aux Sends 3-4/5-6

Aux Sends 3-4/5-6 shift switch
Aux Sends 3-4/5-6 Pre/Post switch
Channel/Mix B source switch

Hi-Mid EQ controls (boost/cut, sweep, bandwidth)

Lo-Mid EQ controls (boost/cut, sweep)

Hi Shelving EQ

Lo Shelving EQ

EQ in switch

Low Cut Filter switch (18dB/octave, 75Hz)

Mix B pan control

Mix B level control

Mix B Split switch

Mix B Source

+22dB Overload LED
-20dB LED
Pan control
Solo switch and LED
Mute switch and LED

Assignment switches (1-2, 3-4, 5-6, 7-8, L-R)

100mm log-taper fader

Figure 11.1 Input strip for the Mackie 8-bus console. (Courtesy of Mackie Designs, www.mackie.com)

capabilities, leaving you with the unpleasant choice of either upgrading or dealing with your present system as best you can. In light of this, it goes without saying that it's always wise to anticipate your future mixing needs when buying a console or mixer.

One way to keep from running out of inputs on your main mixer or console is to use an outboard line mixer (Figures 11.2 and 11.3). These rack-mountable mixers (also known as *submixers*) are often equipped with 16,

Figure 11.2
Mackie LM-3204 4-bus submixer. (Courtesy of Mackie Designs, www.mackie.com)

Figure 11.3
Studio 12R mixer/microphone preamp. (Courtesy of Alesis Studio Electronics, www.alesis.com)

24, or 32 line-level inputs, each having equalization, pan, and effects send capabilities that can be mixed down to either 2 or 4 channels. These channels can then be used to free up a multitude of inputs on your main mixing device.

Mixing via MIDI

Although standard mixing practices still dominate in most multitrack recording studios, MIDI itself can add to the power and flexibility of mixing in a project studio by letting you automate many of the standard mixing functions, directly from your sequencer. That's right! There are no hidden costs and no additional hardware. Truth is, most sequencing packages can give you access to an amazing amount of computerized automation in a simple, easy-to-use environment.

This automation is accomplished by transmitting MIDI channel messages directly to the receiving instruments or devices, in such a way as to provide extensive, real-time control over parameters such as level, panning, and modulation (Figure 11.4).

When combined with power over music voicing, timbre, effects, program changes, etc., MIDI-based mixing and control automation gives the electronic musician a degree of automation over a production that is, in certain respects, unparalleled in audio history.

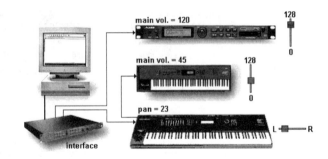

Figure 11.4
System-wide mixing via
MIDI channel messages.

Dynamic Mixing via MIDI Control-Change Messages

The vast majority of electronic instruments allow for dynamic MIDI control over such parameters as velocity (individual note volume), main volume (a device's master output volume), and pan position (of one or more voices). This is accomplished by controlling MIDI voice messages, such as velocity and continuous controller, which can be transmitted as a stream of messages with values that generally range from 0 (minimum) to 127 (maximum). Because these messages can be transmitted over individual MIDI ports and channels, they can be used to control individual voices or a series of grouped voices, so as to form a virtual, software mixer. Because these messages are directly imbedded within the sequence, you can save this automation directly into the sequence file, thereby allowing the mix to be reconfigured whenever the sequence is opened.

Velocity Messages

Although velocity is used to dynamically change the level and expression of individual notes, most sequencers also let you control the overall level of an entire sequenced track or range of notes within a track. For example, let's assume that we have a sequence whose tracks are assigned to a MIDI channel and an associated voice (Figure 11.5). If we want to change the gain of an entire track, portion of a track, or range of tracks, we simply adjust levels by highlighting an entire track or portion of a track and then change the overall velocity levels in that range. This lets you "offset" the values of each note within the selected area by values ranging from 0 to 127 or as a percentage of this range. For example, upon calling up the Change Velocity dialog box, we could "turn up" the defined track volume by simply raising it by a value of 20 (Figure 11.6). The changes could be immediately raised by 20 through the passage and then immediately drop

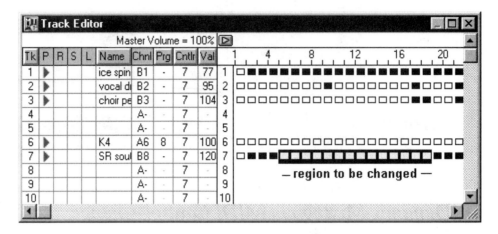

Figure 11.5 Sequencer screen showing region to be gain changed.

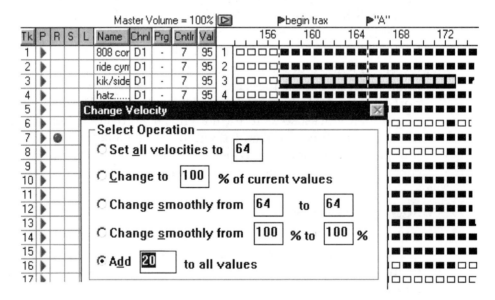

Figure 11.6 Example of how you might enter volume changes into a Change Velocity dialog box. (Courtesy of Passport Designs, Inc., www.passportdesigns.com)

to the original value at the end or they could fade up or down over the range's duration. Depending on the sequencer, the number of gain change possibilities could be fairly vast.

The following examples are a few of the gain-change parameters that might be found in a common sequencing program.

Set velocity:	This gives all notes within the selected range the same MIDI velocity setting.
Adjust velocity:	This proportionately increases or decreases the velocity settings of all notes in the range of measures, using values that range from 0 to 127 or as a percentage of this range.
Fade in:	Smoothly scales the velocity from its initial value to the newly selected value.
Fade out:	Smoothly scales the velocity from its newly selected value to its original value.
Crescendo:	Gradually increases the volume from its initial value to the newly selected velocity settings and back to their original value, essentially adding emphasis to the selected notes.
Limit range:	Limits the velocities to a specified maximum and/or minimum velocity value. This process could be used as a software limiter, in that the levels will not increase beyond a specific level.

Note that, in most cases, changing the actual velocity of a range of notes doesn't have an adverse effect on a sequenced track, other than changing its level; under certain conditions undesirable effects may result. These could occur should velocity levels also affect other sound parameters (i.e., where velocity is used to control such parameters as filter cutoff, reverb mix, or where a note is split between two samples and a velocity change could effect whether a note will sound as a snare hit or as a trash can "SMASH").

Continuous-Controller Messages

As we know from Figure 2.21, the MIDI specification makes provisions for 127 different types of controller messages, of which a few controller numbers have been adopted as *de facto* standards by most manufacturers (i.e., #1—modulation wheel; #4—foot pedal; #7—main volume; #10—pan position). Controller messages 64–95 and 122 have been designated by the MIDI specification as *switches*. These can be set to either 0 (off) or 127 (on), with nothing in between.

Because velocity can be used to control the levels of a sequenced track by changing the volume of each note, certain controller messages can also be used to exert control over the level and dynamic events within a track. One such dynamic event, known as *main volume* (Figure 11.7) uses controller #7 to vary the overall volume of an instrument or the individual voices of a polytimbral instrument.

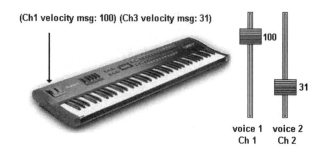

Figure 11.7
Main-volume controller messages within a MIDI system.

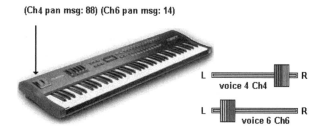

Figure 11.8
Pan-position controller messages in a MIDI system.

In addition, devices that are equipped with stereo outputs or multiple outputs that can be grouped into stereo pairs can use MIDI pan-controller message #10 (Figure 11.8) to pan individual voices or groups of voices from left (value 0) to right (value 127).

A sequencer will often have its own type of interface for accessing MIDI controller messages. In addition, certain MIDI devices may use nonstandard message numbers in order to exert control over dedicated parameters. For these reasons, you should always consult the operating manual for your sequencer, instruments, and other supporting hardware, because they generally include the necessary operating instructions and message data tables.

Mixing a Sequence

Mixing in the MIDI environment isn't only fun, it's often a lesson in understanding the power and flexibility of automated mixing. To begin with, once you have assigned a MIDI channel and MIDI port for each instrument, you can then go about the business of assigning a program change number to an instrument or instrument voice. Once the file is saved, in most cases, reopening the file will automatically reconfigure these settings so that all of the proper sounds will be played back.

The next phase in the process would be to set your overall volume levels for each track. Depending on the sequencer, this mixing interface might take the form of a simulated on-screen mixer (Figure 11.9), it might have volume faders that are part of the main edit screen (Figure 11.10), or it might be called up as a separate window that's dedicated to a controller interface which can be designed from the ground up. In any of these cases, you can begin building your mix by setting the output volumes on each of your instruments to a level that's acceptable to your main mixer or console. (Those of you who are strictly using a soundcard synth don't need to worry about such outboard devices). From a practical standpoint, you might want to set your instrument output levels to full or to some easily identifiable marking, set each of the volume faders that relate to a MIDI instrument to unity gain, and then set the input trim controls on each of these inputs to a level

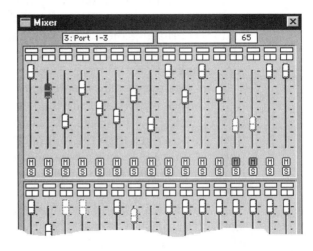

Figure 11.9

An on-screen MIDI mixer window. (Courtesy of Opcode Systems Inc., www.opcode.com)

Tk	P	R	S	L	Name	Chnl	Prg	Cntlr
1	▶				808 cor	D1	-	7
2	▶				ride cym	D1	-	7
3	▶				kik/side	D1	-	7
4	▶				hatz......	D1	-	7
5	▶				Dinodru	D2	-	7
6	▶				Pitz strin	D3	-	7
7	▶	●			Fretless	D4	-	7
8	▶				Kaaas!	D5	-	7
9	▶				Legend	D6	-	7
10	▶				k4 - Sof	A6	13	7
11	▶							7

Figure 11.10   Main volume faders may be part of a sequencer's main edit screen (Courtesy of Passport Designs, Inc., www.passportdesigns.com)

that smoothes the gain differences between each of the instruments and voices. These steps are commonly used in MIDI production because they easily let you reset levels to their standard values should the settings change from mix to mix (as they almost certainly will).

The next step would be to begin playing the sequence and change the fader levels for any instrument or voice until all blend properly into the mix. Once done, you can play back the sequence to see how the overall mix levels hold up over the course of the song. Should the mix need to be changed at any point from its initial settings, most sequencers will let you change the main-volume fader for each track. For example, during the bridge of a song you might want to reduce the volume on several tracks. By entering the automated fader mode (as you might expect, the function names and their usage will vary from one software package to the next), you can manually "ride" the gains, bringing them down at the proper point and then bringing them back up again at the end of the bridge. On playback, the sequencer will generally re-create the fader movements and volume changes in real time.

As was previously mentioned, you also have the option of mixing a defined track region or entire track by directly editing the note velocity levels. Changing the note velocities can sometimes give you more flexibility over gain changes than the faders can, and if your sequencer doesn't support the ability to fade the volume controls smoothly over time, chances are it will let you fade velocity levels over a defined area, so as to create a smooth fade transition.

Certain sequencing packages will also let you group faders into a logical section that can be controlled by simply grabbing a single fader, instead of having to move each track individually. For example, let's say that we have a sequence that contains 6 tracks which are dedicated to percussion. Upon adding a melody line, it quickly becomes obvious that the percussion tracks are too loud. By grouping these tracks together, all we need to do is listen to the sequence and pull any fader in the group down to the proper level and all of the faders in that group will automatically follow.

This last trick calls to mind a very important aspect of most types of production (actually it's the basic tenant of life): Keep it simple! If there's a trick you can use to make your project go more smoothly, use it. For example, most electronic musicians interact with their equipment in a systematic way. They often repeatedly set their tracks and channel assignments up the same way, so the system doesn't have to be repatched or reprogrammed each time you call up a song. To keep life simple, you might want to explore the possibility of creating a basic working template file (Figure 11.11) that has all of the instruments and general assignments already programmed into it.

Tk	P	R	S	L	Name	Chnl	Prg	Cntlr	Val				
								Master Volume = 100% ▷		1	4	8	
1	▶				ASR-10	C1	.	7	127	1	□□□□□□□□□		
2	▶				ASR-10	C2	.	7	127	2	□□□□□□□□□□		
3	▶				ASR-10	C3	.	7	127	3	□□□□□□□□□□		
4	▶				Raven	A2	.	7	127	4	□□□□□□□□□		
5	▶				Rave-o-lution 309	A5	.	7	127	5	□□□□□□□□□		
6	▶	●			K4	A6	2	7	127	6	□□□□□□□□□		
7	▶				Wavestation	B1	.	7	127	7	□□□□□□□□□		
8	▶				Wavestation	B1	.	7	127	8	□□□□□□□□□		
9	▶				X5	A1	.	7	127	9	□□□□□□□□□		
10	▶				Roland (drums)	C6	.	7	127	10	□□□□□□□□□		
11	▶				K4 Perc	A11	.	7	127	11	□□□□□□□□		
12	▶				HR-16 (drums)	A12	34	7	.	12	□□□□□□□		
13	▶					A12	.	7	.	13	□□□□□□		
14	▶				X5 (drums)	B7	55	7	127	14	□□□□		
								7	.	15			

Figure 11.11 A template file (my_setup.mid) can be preprogrammed to include your favorite instruments and general setup assignments.

Now that you've finished the mix, you can master the completed product to the medium of your choice. If the results need to be tweaked, you might consider saving the newly remixed version under a different filename. Because the mix automation is imbedded within the file, saving the various mix versions will let you choose the best version at any time.

When mixing down a sequence in a professional recording studio, some folks prefer not to mix the sequence in the MIDI domain. Instead, they will strip the sequence of its mix-related controller messages (an edit function that's included in most sequencing packages) and then set the main volume and velocities to a reasonably high level. This precaution gives the engineer a greater degree of control over levels at the recording console and multitrack recorder. Such traditional mixing techniques offer an improved signal-to-noise ratio, which results from mixing sound sources that were recorded at the hottest possible signal levels.

You should understand that this isn't a hard and fast rule; it's up to you. Again, should you wish to mix the sequence in the MIDI domain when working at home, you might consider saving the mixed and unmixed versions under different filenames before going into the studio. It never hurts to have more options.

MIDI Remote Controllers

To some, one of the biggest drawbacks to mixing velocity and controller messages from the edit screen of a sequencing program is the lack of a hardware working surface. The ability to put your hands on a tactile surface, grab a fader, and move it into position can be a fast, intuitive, and comforting way of working. Fortunately, such a hardware surface is actually available in the form of a MIDI remote controller.

These cost-effective devices can often give you instant access to such real-time channel messages as main volume, pan messages, and channel solo. They're often designed with either 16 or 24 channels, each having a data slider and switch selectors that can be dedicated to one or more channels and provide control over the output of channel messages.

MIDI remote controllers also have the added bonus of being programmable, allowing them to act as real-time patch editors that can be used with several MIDI instruments and/or devices. This is possible because the controller can be programmed to transmit instrument-specific SysEx data, that can be used to access directly the instrument's CPU for altering patch parameters in real time.

Mixing in Conjunction with a Digital Audio Editor

In modern-day production, it's common for sequencers to work in synchronous tandem with a digital audio editor. Of course, because there are many types of editors on the market, the way that these two media interact varies from system to system.

In its simplest form, a media file containing a General MIDI file, Wave file, and graphics might be simultaneously played back through a single soundcard, with synchronization being loosely handled by the host software itself. In multimedia and electronic music production, however, it's more common to find a situation where the sequencer and digital audio editor are locked together, using MIDI time code as a sync reference.

In such instances, mixing can be handled by each program in its own unique way. As we have seen before, the sequence levels can be balanced from the mixing screen using software faders and other edit commands. The digital audio editor will have it's own mixing interface for changing relative balances and effects processing. By switching between the two programs, an automated mix can be built up and saved to disk.

Other sequencing packages have integrated digital audio capabilities into a single program. This can potentially simplify the mixing process by offering a unified control surface from which the various MIDI and sound file tracks can be mixed.

Finally, a few digital audio workstation systems let you import MIDI files directly into the system. Using this type of interface, you can load a MIDI file into a working project as a track, which can then be mixed as though it were any other audio track.

Digital Remote Controllers

Hardware mixing surfaces (Figure 11.12) have also been designed for use with digital audio editors. Although these devices often have the look and feel of a console (particularly a digital console), they are complex controllers that communicate digital and MIDI data to the system's software. These remote controllers are generally fully automated, often being equipped with moving faders and indicator lights that alert you to channel and system status.

Console Automation

Although MIDI has not generally been implemented into larger, professional console designs, inroads are being made toward integrating MIDI into console and mixer designs that are commonly found in the project studio. Systems such as outboard automation systems, integrated MIDI implementation, and digital consoles have made important strides toward bringing affordable and easily integrated automation packages within the grasp of the electronic musician.

MIDI-Based Automation Systems

One method of achieving low-cost automation makes use of the VCA *(voltage-controlled amplifier)* to exert control over the gain and muting functions of an analog signal. VCAs are used to control a signal level through variations in a DC voltage (generally ranging from 0 to 5 volts) that is applied to the control input of an analog amplifier (Figure 11.13). As the control voltage is increased (in relation to the position of a volume fader), the analog signal is proportionately attenuated. Automation systems commonly encode these control voltages as a series of MIDI SysEx or controller messages that can easily be stored as sequenced MIDI data. On playback, they're used to change the input signal levels in proportion to their original mix positions.

Figure 11.12
The HUI (human user interface) remote controller. (Courtesy of Mackie Designs, www.mackie.com)

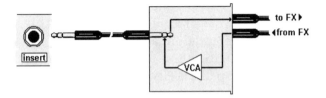

Figure 11.13
Basic diagram of a VCA-based system.

(A) Otto-34 plugs into the direct insert jack of any console to provide VCA-based automation.

(B) Ultrasession automation software for the Mac.

Figure 11.14 VCA-based automation system (Courtesy of Mackie Designs, www.mackie.com)

Devices using this technology often exist as a separate mixing surface that contains faders and switching controls (usually mute/solo buttons) that can be plugged into the insert points of an existing analog console or between the tape deck and console (Figure 11.14). This allows fader and mute functions of an otherwise nonautomated mixer or console to be memorized and then reproduced in real time under MIDI control.

MIDI-Controlled Mixers

Currently, a limited number of audio mixers can respond to MIDI data for the purposes of automation. Such MIDI-controlled mixers can be either analog or digital and often vary in the degree of automation that can be controlled from MIDI. These range from simple MIDI control over mute functions, to the creation of static snapshots for reconfiguring user-defined mixer settings, all the way to complete dynamic control over all mixing functions via MIDI.

Digital Consoles

Developments in digital technology have reduced the price of digital audio mixers and consoles to the point where they are affordable for most professional or well-funded project studios. These devices convert incoming analog signals directly into digital data and distribute and process them throughout the system in the digital domain. At the console's output, the signal can then be decoded back into analog or remain in its digital form for recording to a multitrack or mixdown recorder. Because all audio, routing, and processing functions are digital, full system automation and snapshot recall can be recorded directly into computer memory without any difficulty.

Signal Processors in MIDI Production

As with professional recording technology, MIDI production relies heavily on the use of electronic signal processing for the re-creation of a natural ambiance or for augmenting and modifying audio signals. Modern effects devices offer a wide range of signal processing capabilities (such as reverb, delay, echo, auto pan, flanging, chorusing, equalization, and pitch change). These effects can be varied using control parameters (such as delay time, depth, rate, and filtering), which can be varied to achieve any number of effects. With the implementation of MIDI in many of today's digital signal processors, precise and automated control over effects and their parameters can be achieved.

Effects Automation in MIDI Production

One of the most common ways to automate an effects device from a MIDI sequence or during a live performance is through the use of program-change commands. In the same way that electronic instruments can store sound patches into a memory location register for later recall, most modern effects devices will let you store program patches into memory, where they can be automatically recalled using a program-change command.

The use of program-change commands (and occasionally continuous-controller messages) makes it possible for signal processing patches and parameters to be altered during the playback of a MIDI sequence. In most cases, a program change that relates to the desired effects patch can be inserted at the beginning of a sequenced track, so that the proper effect will be automatically called up once the sequence is opened.

Dynamic Effects Editing via MIDI

In addition to real-time program changes, the dynamic editing of effects parameters is often possible by transmitting real-time SysEx messages, letting you have real-time control over preset effects parameters (i.e., program type, reverb time, equalization, or chorus depth). This type of control is often handled from a dedicated remote controller, MIDI remote controller, or software patch editor through the transmission of SysEx messages.

By using a software patch editor, it's often possible to edit and fine-tune effects parameters from an on-screen graphic interface that directly represents the device's control parameter settings. Once the desired effect or effects have been assembled and fine-tuned, these settings can be saved to disk and/or the devices own memory register for recall at a later time.

Once a device's bank of preset locations has been filled, a software patch editor can receive a full SysEx data dump of all the patches, allowing multiple preset banks to be stored and recalled for future use. Many of them will also let you organize effects patches into groups (in a library fashion) according to effects or any other category. When it comes to SysEx data dumps, don't forget that most sequencers will let you load and retransmit SysEx patch dumps directly from the sequencer. This powerful feature lets you reconfigure your effects devices without having to leave the sequencing program.

As with most instrument patches, effects patches can be acquired from a number of sources. Among these are patch books (containing written patch data for manual entry), patch data cards (ROM cards, cartridges from the manufacturers or third-party developers), data disks (computer files containing patch data from manufacturers or third-party developers), and, of course, the web (which contains more patch-related data dumps than you can shake a hundred synths at).

MIDI Signal Processing and Effects Devices

Over the years, the number of signal processing and effects devices that implement MIDI into their control structure has increased to the point that it's almost a present-day given. This is due to the power and increased flexibility that effects can bring to MIDI production, as well as the increased editing and control capabilities that MIDI can bring to the field of signal processing.

By far the most commonly found MIDI controlled and programmable devices are digital effects processors. These devices are commonly used to either augment or modify a signal, so as to improve its production value within a mix. Such a processor operates by converting analog signals into corresponding digital data. This data is then processed via user-defined algorithms (computer programs for performing complex calculations on digital data according to a predetermined pattern) that can be varied to customize an effect to best suit a sound or music mix. Many of these effects devices are capable of processing an audio signal using more than one algorithm, allowing the processor to produce a number of simultaneous effects.

Musical Instrument Processors

Various effects devices are also being designed expressly for use with musical instruments (such as keyboards and electric guitar systems). These devices offer a wide range of effects that can be designed and organized into banks that best suit their particular instrument. In addition to the effects box itself, MIDI program-change "stomp boxes" let you select the effect that you want on the fly via program-change messages.

In summary, all of the devices in this chapter can work in tandem to give your mix a degree of automation and connectivity that is unparalleled in music history—and at prices that are within the reach of practically anyone who is willing to make use of the power and potential of what MIDI has to offer. It's up to you now. Dive into the deep end and have fun!

A

THE MIDI 1.0 SPECIFICATION*

Introduction

MIDI is the acronym for Musical Instrument Digital Interface. MIDI enables synthesizers, sequencers, home computers, rhythm machines, etc., to be interconnected through a standard interface.

Each MIDI-equipped instrument usually contains a receiver and a transmitter. Some instruments may contain only a receiver or transmitter. The receiver receives messages in MIDI format and executes MIDI commands. It consists of an opto-isolator, Universal Asynchronous Receiver/Transmitter (UART), and other hardware needed to perform the intended functions. The transmitter originates messages in MIDI format and transmits them by way of a UART and line driver.

The MIDI standard hardware and data format are defined in this specification.

Conventions

Status and data bytes given in Tables I through VI are given in binary. Numbers followed by an "H" are in hexadecimal. All other numbers are in decimal.

* MIDI Manufacturers Association, 5316 West 57th Street, Los Angeles, CA, 90056; www.midi.org.

Hardware

The interface operates at 31.25 (+/– 1%) kbaud, asynchronous, with a start bit, 8 data bits (D0 to D7), and a stop bit. This makes a total of 10 bits for a period of 320 microseconds per serial byte.

Circuit: 5-mA current loop type. Logical 0 is current ON. One output shall drive one and only one input. The receiver shall be opto-isolated and require less than 5 mA to turn on. Sharp PC-900 and HP 6N138 opto-isolators have been found acceptable. Other high-speed opto-isolators may be satisfactory. Rise and fall times should be less than 2 microseconds.

Connectors: DIN 5-pin (180°) female panel mount receptacle. An example is the SWITCHCRAFT 57 GB5F. The connectors shall be labeled "MIDI IN" and "MIDI OUT." Note that pins 1 and 3 are not used, and should be left unconnected in the receiver and transmitter (Figure A.1).

Cables shall have a maximum length of 50 feet (15 meters), and shall be terminated on each end by a corresponding 5-pin DIN male plug, such as the SWITCHCRAFT 05GM5M. The cable shall be shielded twisted pair, with the shield connected to pin 2 at both ends.

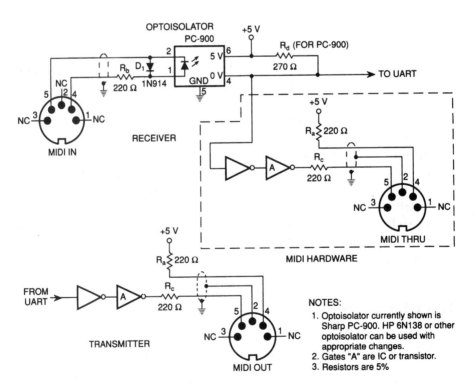

Figure A.1 Standard hardware configuration for MIDI in, out, and thru ports.

A "MIDI THRU" output may be provided if needed, which provides a direct copy of data coming in MIDI IN. For very long chain lengths (more than three instruments), higher speed opto-isolators must be used to avoid additive rise/fall time errors that affect the pulse width duty cycle.

Data Format

All MIDI communication is achieved through multibyte "messages" consisting of one status byte followed by one or two data bytes, except for real-time and exclusive messages (see below).

Message Types

Messages are divided into two main categories: channel and system.

Channel

Channel messages contain a 4-bit number in the status byte that addresses the message specifically to 1 of 16 channels. These messages are thereby intended for any units in a system whose channel number matches the channel number encoded into the status byte.

There are two types of channel messages: voice and mode. To control the instrument's voices, *voice messages* are sent over the voice channels. To define the instrument's response to voice messages, *mode messages* are sent over the instrument's basic channel.

System

System messages are not encoded with channel numbers. There are three types of system messages: common, real-time, and exclusive.

Common: Common messages are intended for all units in a system.

Real-Time: Real-time messages are intended for all units in a system. They contain status bytes only—no data bytes. Real-time messages may be sent at any time—even between bytes of a message that has a different status. In such cases, the real-time message is either ignored or acted on, after which the receiving process resumes under the previous status.

Exclusive: Exclusive messages can contain any number of data bytes, and are terminated by an end-of-exclusive (EOX) or any other status byte. These messages include a manufacturer's identification (ID) code. If the receiver does not recognize the ID code, it should ignore the ensuing data.

So that other users can fully access MIDI instruments, manufacturers should publish the format of data following their ID code. Only the manufacturer can update the format following their ID.

Data Types

Status Bytes

Status bytes are 8-bit binary numbers in which the most significant bit (MSB) is set (binary 1). Status bytes serve to identify the message type; that is, the purpose of the data bytes that follow the status byte.

Except for real-time messages, new status bytes will always command the receiver to adopt their status, even if the new status is received before the last message was completed.

Running status: For voice and mode messages only, when a status byte is received and processed, the receiver will remain in that status until a different status byte is received. Therefore, if the same status byte would be repeated, it may (optionally) be omitted so that only the correct number of data bytes need be sent. Under running status, then, a complete message need only consist of specified data bytes sent in the specified order.

The running status feature is especially useful for communicating long strings of note-on/note-off messages, where "Note On with Velocity of 0" is used for note-off. (A separate note-off status byte is also available.) Running status will be stopped when any other status byte intervenes, except that real-time messages will only interrupt the running status temporarily.

Unimplemented status: Any status bytes received for functions that the receiver has not implemented should be ignored, and subsequent data bytes ignored.

Undefined status: Undefined status bytes must not be used. Care should be taken to prevent illegal messages from being sent during power-up or power-down. If undefined status bytes are received, they should be ignored, as should subsequent data bytes.

Data Bytes

Following the status byte, there are (except for real-time messages) one or two data bytes which carry the content of the message. Data bytes are 8-bit binary numbers in which the MSB is reset (binary 0). The number and range of data bytes that must follow each status byte are specified in the tables that follow. For each status byte the correct number of data bytes must always be sent. Inside the receiver, action on the message should wait until all data bytes required under the current status are received. Receivers should ignore data bytes that have not been properly preceded by a valid status byte (with the exception of "Running Status," see above).

Channel Modes

Synthesizers contain sound generation elements called *voices*. Voice assignment is the algorithmic process of routing note-on/note-off data from the keyboard to the voices so that the musical notes are correctly played with accurate timing.

When MIDI is implemented, the relationship between the 16 available MIDI channels and the synthesizer's voice assignment must be defined. Several mode messages are available for this purpose (see Table III). They are Omni (On/Off), Poly, and Mono. Poly and Mono are mutually exclusive, that is, Poly Select disables Mono, and vice versa. Omni, when on, enables the receiver to receive voice messages in all voice channels without discrimination. When Omni is off, the receiver will accept voice messages from only the selected voice channel(s). Mono, when on, restricts the assignment of voices to just one voice per voice channel (monophonic). When Mono is off (= Poly on), any number of voices may be allocated by the receiver's normal voice assignment algorithm (polyphonic).

For a receiver assigned to basic channel N, the four possible modes arising from the two mode messages are as follows:

Mode	Omni		
1	On	Poly	Voice messages are received from all voice channels and assigned to voices polyphonically.
2	On	Mono	Voice messages are received from all voice channels, and control only one voice, monophonically.
3	Off	Poly	Voice messages are received in voice channel N only and are assigned to voices polyphonically.
4	Off	Mono	Voice messages are received in voice channels N thru $N + M - 1$, and assigned monophonically to voices 1 thru M, respectively. The number of voices M is specified by the third byte of the Mono Mode Message specified by the third byte of the Mono Mode Message.

Four modes are applied to transmitters (also assigned to basic channel N). Transmitters with no channel selection capability will normally transmit on basic channel 1 (N=0).

Mode	Omni		
1	On	Poly	All voice messages are transmitted in Channel N.
2	On	Mono	Voice messages for one voice are sent in Channel N.
3	Off	Poly	Voice messages for all voices are sent in Channel N.
4	Off	Mono	Voice messages for voices 1 thru M are transmitted in Voice Channels N thru N + M – 1, respectively. (Single voice per channel.)

A MIDI receiver or transmitter can operate under one and only one mode at a time. Usually the receiver and transmitter will be in the same mode. If a mode cannot be honored by the receiver, it may ignore the message (and any subsequent data bytes), or it may switch to an alternate mode (usually Mode 1, Omni On/Poly).

Mode messages will be recognized by a receiver only when sent in the basic channel to which the receiver has been assigned, regardless of the current mode. Voice messages may be received in the basic channel and in other channels (which are all called voice channels), which are related specifically to the basic channel by the rules above, depending on which mode has been selected.

A MIDI receiver may be assigned to one or more basic channels by default or by user control. For example, an eight-voice synthesizer might be assigned to basic channel 1 on power-up. The user could then switch the instrument to be configured as two four-voice synthesizers, each assigned to its own basic channel. Separate mode messages would then be sent to each four-voice synthesizer, just as if they were physically separate instruments.

Power-Up Default Conditions

On power-up all instruments should default to Mode #1. Except for note-on/note-off status, all voice messages should be disabled. Spurious or undefined transmissions must be suppressed.

Table I Summary of Status Bytes

Status (D7–D0)	# of Data Bytes	Description
		Channel Voice Messages
1000nnnn	2	Note Off event
1001nnnn	2	Note On event (velocity=0: Note Off)
1010nnnn	2	Polyphonic key pressure/after touch
1011nnnn	2	Control change
1100nnnn	1	Program change
1101nnnn	1	Channel pressure/after touch
1110nnnn	2	Pitch bend change
		Channel Mode Messages
1011nnnn	2	Selects Channel Mode
		System Messages
11110000	*****	System Exclusive
11110sss	0 to 2	System Common
11111ttt	0	System Real Time

Notes:

nnnn: N – 1, where N = Channel #,
i.e., 0000 is Channel 1.
0001 is Channel 2.
-
-
-
1111 is Channel 16.

*****: Oiiiiiii, data, ..., EOX

iiiiiii: Identification

sss: 1 to 7

ttt: 0 to 7

Table II   Channel Voice Messages

Status	Data Bytes	Description
1000nnnn	Okkkkkk	Note Off (see notes 1–4)
	0vvvvvvv	vvvvvvv: note off velocity
1001nnnn	Okkkkkkk	Note On (see notes 1–4)
	0vvvvvvv	vvvvvvv-0: velocity
		vvvvvvv = 0: note off
1010nnnn	Okkkkkk	Polyphonic Key Pressure (After Touch)
	0vvvvvv	vvvvvvv: pressure value
1011nnnn	0ccccccc	Control Change
	0vvvvvvv	ccccccc: control # (0–121) (see notes 5–8)
		vvvvvvv: control value
		ccccccc = 122 thru 127: Reserved. (See Table III)
1100nnnn	0ppppppp	Program Change
		ppppppp: program number (0–127)
1101nnnn	0vvvvvvv	Channel Pressure (After Touch)
		vvvvvvv: pressure value
1110nnnn	0vvvvvvv	Pitch Bend Change LSB (see note 10)
	0vvvvvvv	Pitch Bend Change MSB

Notes:

1. nnnn: Voice Channel # (1–16, coded as defined in Table I notes)

2. kkkkkkk: note # (0–127)
   kkkkkkk = 60: Middle C of keyboard

0	12	24	36	48	60	72	84	96	108	120	127
		ac	c	c	c	c	c	c	c		

   $\vdash$————— piano range —————$\dashv$

3. vvvvvvv: key velocity (A logarithmic scale would be advisable.)

0	1			64				127
off	ppp	pp	p	mp	mf	f	fff	fff

   vvvvvvv= 64: in case of no velocity sensors
   vvvvvvv = 0: Note Off, with velocity = 64

4. Any Note On message sent should be balanced by sending a Note Off message for that note in that channel at some later time.

5. ccccccc: control number

ccccccc	**Description**
0	Continuous Controller 0 MSB
1	Continuous Controller 1 MSB (MODULATION BENDER)
2	Continuous Controller 2 MSB
3	Continuous Controller 3 MSB
4-31	Continuous Controllers 4-31 MSB
32	Continuous Controller 0 LSB
33	Continuous Controller 1 LSB (MODULATION BENDER)
34	Continuous Controller 2 LSB
35	Continuous Controller 3 LSB
36-63	Continuous Controllers 4-31 LSB
64-95	Switches (On/Off)
96-121	Undefined
122-127	Reserved for Channel Mode messages (see Table III).

6. All controllers are specifically defined by agreement of the MIDI Manufacturers Association (MMA) and the Japan MIDI Standards Committee (JMSC). Manufacturers can request through the MMA or JMSC that logical controllers be assigned to physical ones as necessary. The controller allocation table must be provided in the user's operation manual.

7. Continuous controllers are divided into Most Significant and Least Significant Bytes. If only seven bits of resolution are needed for any particular controllers, only the MSB is sent. It is not necessary to send the LSB. If more resolution is needed, then both are sent, first the MSB, then the LSB. If only the LSB has changed in value, the LSB may be sent without resending the MSB.

8. vvvvvvv: control value (MSB)
   (for controllers)

   0 ————————————————————————————————127
   min   max

   (for switches)

   0 - 127
   off   on

   Numbers 1 through 126, inclusive, are ignored.

9. Any messages (e.g., Note On), which are sent successively under the same status, can be sent without a status byte until a different status byte is needed.

10. Sensitivity of the pitch bender is selected in the receiver. Center position value (no pitch change) is 2000H, which would be transmitted EnH-00H-40H.

Table III   Channel Mode Messages

Status	Data Bytes	Description
1011nnnn	0ccccccc 0vvvvvvv	Mode Messages
		ccccccc = 122: Local Control
		vvvvvvv = 0, Local Control Off
		vvvvvvv = 127, Local Control On
		ccccccc = 123: All Notes Off
		vvvvvvv = 0
		ccccccc = 124: Omni Mode Off (All Notes Off)
		vvvvvvv = 0
		ccccccc = 125: Omni Mode On (All Notes Off)
		vvvvvvv = 0
		ccccccc = 126: Mono Mode On (Poly Mode Off), (All Notes Off)
		vvvvvvv = M, where M is the number of channels.
		vvvvvvv = 0, the number of channels equals the number of voices in the receiver.
		ccccccc = 127: Poly Mode On (Mono Mode Off)
		vvvvvvv = 0 (All Notes Off)

Notes:
1. nnnn: basic channel # (1–16, coded as defined in Table I)
2. Messages 123 thru 127 function as All Notes Off messages. They will turn off all voices controlled by the assigned basic channel. Except for message 123, All Notes Off, they should not be sent periodically, but only for a specific purpose. In no case should they be used in lieu of Note Off commands to turn off notes which have been previously turned on. Therefore, any All Notes Off command (123–127) may be ignored by receiver with no possibility of notes staying on, since any

Note On command must have a corresponding specific Note Off command.

3. Control Change #122, Local Control, is optionally used to interrupt the internal control path between the keyboard, for example, and the sound generating circuitry. If 0 (Local Off message) is received, the path is disconnected: The keyboard data goes only to MIDI and the sound generating circuitry is controlled only by incoming MIDI data. If a 7FH (Local On message) is received, normal operation is restored.

4. The third byte of "Mono" specifies the number of channels in which Monophonic Voice messages are to be sent. This number, "M," is a number between 1 and 16. The channel(s) being used, then, will be the current basic channel (=N) thru N + M − 1 up to a maximum of 16. If M=0, this is a special case directing the receiver to assign all its voices, one per channel, from the basic channel N through 16.

Table IV System Common Messages

Status	Data Bytes	Description
11110001		Undefined
11110010		Song Position Pointer
	0lllllll	lllllll: (Least significant)
	0hhhhhhh	hhhhhhh: (Most significant)
11110011	0sssssss	Song Select
		sssssss: Song #
11110100		Undefined
11110101		Undefined
11110110	none	Tune Request
11110111	none	EOX: "End of System Exclusive" flag

1. Song Position Pointer: Is an internal register which holds the number of MIDI beats (1 beat = 6 MIDI clocks) since the start of the song. Normally it is set to 0 when the START switch is pressed, which starts sequence playback. It then increments with every sixth MIDI clock receipt, until STOP is pressed. If CONTINUE is pressed, it continues to increment. It can be arbitrarily preset (to a resolution of 1 beat) by the SONG POSITION POINTER message.

2. Song Select: Specifies which song or sequence is to be played upon receipt of a start (real-time) message.

3. Tune Request: Used with analog synthesizers to request them to tune their oscillators.

4. EOX: Used as a flag to indicate the end of a System Exclusive transmission (see Table VI).

Table V System Real-Time Messages

Status	Data Bytes	Description
11111000		Timing Clock
11111001		Undefined
11111010		Start
11111011		Continue
11111100		Stop
11111101		Undefined
11111110		Active Sensing
11111111		System Reset

Notes:

1. The system real-time messages are for synchronizing all of the system in real time.

2. The system real-time messages can be sent at anytime. Any messages which consist of two or more bytes may be split to insert real-time messages.

3. Timing clock (F8H): The system is synchronized with this clock, which is sent at a rate of 24 clocks/quarter note.

4. Start (from the beginning of song) (FAH): This byte is immediately sent when the PLAY switch on the master (e.g., sequencer or rhythm unit) is pressed.

5. Continue (FBH): This is sent when the CONTINUE switch is hit. A sequence will continue at the time of the next clock.

6. Stop (FCH): This byte is immediately sent when the STOP switch is hit. It will stop the sequence.

7. Active Sensing (FEH): Use of this message is optional, for either receivers or transmitters. This is a "dummy" status byte that is sent

every 300 ms (max), whenever there is no other activity on MIDI. The receiver will operate normally if it never receives FEH. Otherwise, if FEH is ever received, the receiver will expect to receive FEH or a transmission of any type every 300 ms (max). If a period of 300 ms passes with no activity, the receiver will turn off the voices and return to normal operation.

8. System Reset (FFH): This message initializes all of the system to the condition of just having turned on power. The system reset message should be used sparingly, preferably under manual command only. In particular, it should not be sent automatically on power up.

Table VI System Exclusive Messages

Status	Data Bytes	Description
11110000		Bulk dump, etc.
	0iiiiiii	iiiiiii: identification
	*	
	(0*******)	
	*	Any number of bytes may be sent here, for any purpose, as long as they all have a zero in the most significant bit.
	(0*******)	
	*	
	11110111	EOX: "End of System Exclusive"

Notes:
1. iiiiiii: identification ID (0–127)
2. All bytes between the system exclusive status byte and EOX or the next status byte must have zeros in the MSB.
3. The ID number can be obtained from the MMA or JMSC.
4. In no case should other status or data bytes (except real-time) be interleaved with system exclusive, regardless of whether or not the ID code is recognized.
5. EOX or any other status byte, except real-time, will terminate a system exclusive message, and should be sent immediately at its conclusion.

"MIDI from the Source ..."

The *Complete MIDI 1.0 Detailed Specification* is the definitive book on MIDI for developers, musicians, hobbyists, technicians, or just about anyone who wants to know about the inner workings of MIDI. The new edition is fully updated by the MMA and provides details of all approved MIDI messages and recommended practices.

Various books and articles are also available that describe the MIDI 1.0 specification, available in print and electronically, but the recent additions or changes to the specification are found only in the official MMA document.

The *Complete MIDI 1.0 Detailed Specification* describes all of the approved MIDI messages and recommended practices and includes sections on the following topics: MIDI 1.0, General MIDI, standard MIDI files, MIDI Show Control (NEW! Version 1.1), MIDI Machine Control, and MIDI Time Code. The MMA document also includes a tutorial on music synthesis and MIDI for those who may be unfamiliar with musical instrument design.

Document Revision History

Although the MIDI specification is still called "MIDI 1.0" there have been many enhancements and updates made since the original specification was written in 1984. Besides the addition of new MIDI messages such as the MIDI Machine Control and MIDI Show Control messages, there have also been improvements to the "basic" protocol, adding features such as Bank Select, All Sound Off, and many other new controller commands.

Until 1995 there were five separate documents covering the basic MIDI specification, the additions (MSC and MMC), plus standard MIDI files and General MIDI. The "95.1" version—published in January 1995—compiled the latest versions of these documents. The version of the basic MIDI specification (called the Detailed Specification) was version 4.2 prior to that time, which was itself a compilation of the "Detailed Specification v4.2" document and the "4.2 Addendum." Versions 95.1 and 95.2 integrated the existing documents and fixed some minor errors in the various documents.

Ordering Information

As of this writing, The *Complete MIDI 1.0 Detailed Specification* can be purchased for US$49.95 plus postage by writing the MMA directly at P.O. Box 3173, La Habra, CA 90632-3173, www.midi.org.

B

THE MIDI IMPLEMENTATION CHART

It isn't always necessary for a MIDI device to transmit or receive every type of MIDI message defined by the MIDI specification, because certain messages might not relate to the device's function. For example, a synth module is only required to respond to note-on/note-off messages. Because the device does not have a built-in keyboard, there's no need to transmit these messages. Other devices might limit certain MIDI messages due to factors such as design limitations or cost effectiveness. For example, no amount of keyboard banging on a velocity-sensitive keyboard controller will vary the output volume on a synth that does not respond to velocity messages.

To ensure that two or more MIDI devices will be able to communicate MIDI events effectively, the MMA (MIDI Manufacturers Association, www.midi.org) and the JMSC (Japan MIDI Standard Committee) have devised a *MIDI Implementation Chart* (Figure B.1) that lets the reader look at all the MIDI capabilities of a specific MIDI device at a single glance using a standardized printed format.

From the user's standpoint, it is always wise to compare implementation charts with other devices within the existing MIDI system when considering a new piece of equipment. This will ensure that the device will recognize existing messages and add to your system's current capabilities.

Guidelines for Using the Chart

The MMA specifies that the MIDI implementation chart (Figure B.1) be printed the same size using a standardized spreadsheet format consisting of 4 columns by 12 rows. The first column lists the MIDI function in question. The second lists information relating to whether (or how) the device transmits this function's data. The third lists whether (or how) the device recognizes (receives) this data, and the final column is used for additional remarks by the manufacturer.

Despite efforts at standardization, slight inconsistencies within the chart's specifications allow for variations in the symbols, abbreviations, spelling, etc., that can be used by different manufacturers. The following guidelines provide a basic understanding of these differences.

- In general, the symbol 0 is used to indicate that a MIDI function is implemented, while an X is used to show that the function is not implemented. However, some charts may use an X to equal a yes and an 0 to equal a no. This is usually indicated within a key at the lower right-hand corner of the chart.
- OX or an asterisk (*) is used to indicate a selectable function. Further information on the range or type of selectability will be placed within the remarks column.
- MIDI modes are listed as follows: mode 1 (Omni on, Poly), mode 2 (Omni on, Mono), mode 3 (Omni off, Poly), and mode 4 (Omni off, Mono). The modes will often be listed at the bottom of the chart. Occasionally abbreviations of these modes (i.e., Omni on/off, Omni on, or Poly) may be used by a manufacturer.

Detailed Explanation of the Chart

The following paragraphs provide a detailed explanation of the various functions and their related categories that are found within the chart.

Header

The *header* provides the user with the model number, brief description, date, and version number of the device.

Basic Channel

Basic channel indicates which MIDI channels are used by the device to transmit and receive data. The subheadings for this function are default and changed.

	Function •••	Transmitted	Recognized	Remarks
Basic Channel	Default Changed	1 – 16 1 – 16	1 – 16 1 – 16	memorized
Mode	Default Messages Alterd	Mode 3, 4 OMNI OFF, MONO POLY * * * * * * * *	Mode 3 × 	memorized
Note Number	True Voice	0 – 127 * * * * * * * * *	0 – 127 12 – 108	
Velocity	Note ON Note OFF	○ v = 1 – 127 × 9n v = 0	○ v = 1 – 127 ×	
After Touch	Key's Ch's	× ×	× ×	
Pitch Bender		○	○ 0 – 24 semitone	
Control Change	1 2 – 5 6 7 8 – 15 16 17 – 37 38 39 – 63 64 65 – 80 81 82 – 99 100 – 101 102 – 120 121	○ × * * ○ × × × * * × × × × × * * (0) × ○	○ × * * ○ × ○ × × × ○ × ○ × * * (0) × ○	Modulation Data Entry MSB Volume General Purpose Control-1 Data Entry LSB Hold 1 General Purpose Control-1 RPC LSB, MSB Reset All Controllers
Prog Change	True #	○ 0 – 127 * * * * * * * * *	○ 0 – 127 0 – 127	
System Exclusive		○	○	
System Common	Song Pos Song Sel Tune	× × ×	× × ×	
System Real Time	Clock Commands	× ×	× ×	
Aux Message	Local ON/OFF All Notes OFF Active Sense Reset	× × ○ ×	× ○ ○ ○	
Notes		*Control Change messages from 0 to 95 which are recognized through Control channel are transmitted throgh all the channels which are used in Branches. Howëver, General Purpose Control -1 and General Purpose Control – 6 are converted into the same functions as the FC-100 EV-5 assign and the FC-100 Switch assign in the System Setup, and are transmitted. * *RPC = Registered Parameter Control Number RPC # 0 : Bender Range The value of parameter is to be determined by entering data.		

Mode 1 : OMNI ON, POLY   Mode 2 : OMNI ON, MONO   ○ : Yes
Mode 3 : OMNI OFF, POLY   Mode 4 : OMNI OFF, MONO   × : No

Figure B.1   Example of a MIDI Implementation Chart.

- *Default:* This indicates which MIDI channel is in use when the device is first turned on.
- *Changed:* This indicates which of the MIDI channels can be addressed after the device is first turned on.

Mode

Mode indicates which of the MIDI modes may be used by the device. The subheadings for this function are default, messages, and altered.

- *Default:* This indicates which of the four MIDI modes is active when the device is first turned on.
- *Messages:* This describes which of the four MIDI modes can be transmitted or recognized by the device.
- *Altered:* This refers to mode messages that cannot be recognized by the device. It may be followed by a description of the mode that the device automatically enters into upon receiving a request message for an unavailable mode.

Note Number

The transmitted *note number* indicates the range of MIDI note numbers that are transmitted by a device. The maximum possible range spans from 0 to 127, while 21 to 108 corresponds to the 88 keys of an extended keyboard controller. Should the note number be greater than the actual number of keys on a keyboard device, a key transposition feature is indicated.

The recognized note number indicates the range of MIDI note numbers that can be recognized by a device. MIDI notes that are out of this range are ignored by this device. A second note number range, known as *true voice*, indicates the number of notes the device can actually play. Recognized notes that are out of the actual voice range are transposed up or down in octaves until they fall within this range.

Velocity

This category indicates whether the device is capable of transmitting or receiving attack and release velocity messages. The subheadings for this function are *note on* and *note off*.

- *Note on:* This indicates if the device is capable of transmitting and responding to variable-velocity (attack) messages. Not all dynamically controllable devices respond to the full velocity range (1–127). Some devices, such as drum machines, respond to a finite number of velocity steps.

- *Note off:* This indicates whether the device is capable of transmitting and responding to variable release velocity messages. Many devices use a message (note-on velocity = 0) to indicate a note-off condition. This is often indicated in the chart by 9NH v=O or $9n 00, which is the hexadecimal equivalent for this message.

After Touch

After touch indicates how pressure data is transmitted or received. The subheadings for this function are key's and ch's.

- *Key's:* This indicates if the device will transmit or receive independent polyphonic-pressure messages for each key.
- *Ch's:* This indicates whether the device is capable of transmitting or receiving channel-pressure changes (a common after-touch mode, providing one pressure value for an entire MIDI channel).

Pitch Bender

Pitch bender indicates if the device is capable of transmitting or receiving pitch-bend information. If so, the remarks column will often give information as to the pitch-bend range and resolution.

Control Change

Control change indicates whether the device is capable of transmitting or receiving continuous-controller messages. The chart will often list which of these messages are supported in addition to providing a detailed breakdown of their parameters within the remarks column.

Program Change

This category indicates if the device is capable of transmitting or receiving *program-change messages.* True # indicates the message numbers that are actually supported by the device's program-change buttons.

System Exclusive

This indicates if the device is capable of transmitting and receiving *system exclusive data.* The remarks column will often give general information as to which type of SysEx data is supported. However, more detailed data will generally be provided within the device's manual.

System Common

This indicates whether the device is capable of transmitting or receiving the different types of *system common messages*, such as SPP, MIDI time code, song select, and tune-request messages.

System Real Time

This category indicates whether the device can transmit or receive *system real-time messages*. The subheadings for this function are *clock* and *commands*.

- *Clock:* This refers to the device's ability to receive or transmit MIDI clock messages. A device that can transmit MIDI clock messages may be used to provide master timing information within a MIDI system, while a device capable of receiving clock data may only be slaved to other MIDI devices.
- *Commands:* This indicates whether the device is capable of transmitting or responding to start, stop, and continue messages.

Auxiliary Messages

This indicates if a device is capable of transmitting or receiving local control-on/control-off, all notes-off, active-sensing, and system-reset messages.

Notes

This area is used by the manufacturer to comment on any function or implementation particular to the specific MIDI device.

C

CONTINUED EDUCATION

As the equipment used by electronic musicians places an ever-increasing emphasis on technology, education must play a greater role in the understanding of basic industry skills. Education can take many forms, ranging from a formal education to simply keeping abreast of industry directions from the many industry magazines.

Beyond getting a hands-on education by playing with all of the tools and toys, one of the best ways to get a better handle on the pulse of this industry is to *read*. Read all the magazines, books, and articles that you can get your hands on. In addition to reading, many schools offer courses in electronic music production. If this approach works for you, I'd recommend that you carefully check the institution and department out before potentially laying down your hard-earned cash to further your education in electronic music technology.

An expanded and continually updated listing of current music industry-related magazines, trade shows, and press-related information can be obtained by contacting either the mediamanager site (www.mediamanager.com) or the Music Yellow Pages (www.musicyellowpages.com).

The Web

For those MIDI professionals and enthusiasts out there who don't have access to the World Wide Web, I really encourage you to get your hands on a modem (28.8 kbps or faster), find an Internet provider that fits your needs and budget, and start surfing the "Net."

One of the best ways to begin surfing the web is to log on to a popular search engine, such as Yahoo (www.yahoo.com) or AltaVista (www.altavista.com). Or at the search prompt, simply type "midi" and watch the world of information—such as SysEx patches, standard MIDI files, product reviews and tons more—open up before your very eyes.

If you want to access a company or organization directly (in order to get product information, specs, pricing information, or even to download product manuals electronically), you can often enter the name (using lowercase letters) in the following form: www.companyname.com or www.organizationname.org. For example, in order to log onto Yamaha's home site, you simply type in www.yamaha.com.

Users can also use the web to communicate using public and private forums (allowing the users to voice their comments and ideas to other users and manufacturers) in addition to contacting the company directly using electronic mail services. . . . Happy surfing!

INDEX

X